Texts and Monographs in Physics

Springer
Berlin
Heidelberg
New York
Barcelona
Budapest
Hong Kong
London
Milan
Paris
Santa Clara
Singapore
Tokyo

Texts and Monographs in Physics

From Microphysics to Macrophysics I + II Methods and Applications of Statistical Physics By R. Balian

Variational Methods in Mathematical Physics A Unified Approach By P. Blanchard and E. Brüning

Quantum Mechanics: Foundations and Applications 3rd enlarged edition By A. Böhm

The Early Universe Facts and Fiction 3rd corrected and enlarged edition By G. Börner

Operator Algebras and Quantum Statistical Mechanics I + II 2nd edition By O. Bratteli and D. W. Robinson

Geometry of the Standard Model of Elementary Particles By A. Derdzinski

Scattering Theory of Classical and Quantum *N*-Particle Systems By J. Dereziński and C. Gérard

Effective Lagrangians for the Standard Model By A. Dobado, A. Gómez-Nicola, A. L. Maroto and J. R. Peláez

Quantum The Quantum Theory of Particles, Fields, and Cosmology By E. Elbaz

Quantum Relativity A Synthesis of the Ideas of Einstein and Heisenberg By D. R. Finkelstein

Quantum Mechanics I + II By A. Galindo and P. Pascual

The Elements of Mechanics By G. Gallavotti

Local Quantum Physics Fields, Particles, Algebras 2nd revised and enlarged edition By R. Haag

Supersymmetric Methods in Quantum and Statistical Physics By G. Junker

***CP* Violation Without Strangeness** Electric Dipole Moments of Particles, Atoms, and Molecules By I. B. Khriplovich and S. K. Lamoreaux

Inverse Schrödinger Scattering in Three Dimensions By R. G. Newton

Scattering Theory of Waves and Particles 2nd edition By R. G. Newton

Quantum Entropy and Its Use By M. Ohya and D. Petz

Generalized Coherent States and Their Applications By A. Perelomov

Essential Relativity Special, General, and Cosmological Revised 2nd edition By W. Rindler

Path Integral Approach to Quantum Physics An Introduction 2nd printing By G. Roepstorff

Finite Quantum Electrodynamics The Causal Approach 2nd edition By G. Scharf

From Electrostatics to Optics A Concise Electrodynamics Course By G. Scharf

The Mechanics and Thermodynamics of Continuous Media By M. Šilhavý

Large Scale Dynamics of Interacting Particles By H. Spohn

The Theory of Quark and Gluon Interactions 2nd completely revised and enlarged edition By F. J. Ynduráin

Relativistic Quantum Mechanics and Introduction to Field Theory By F. J. Ynduráin

Iosif B. Khriplovich
Steve K. Lamoreaux

CP Violation Without Strangeness

Electric Dipole Moments of Particles, Atoms, and Molecules

With 36 Figures

Springer

Iosif B. Khriplovich

Budker Institute of Nuclear Physics
630090 Novosibirsk, Russia

Steve K. Lamoreaux

University of California
Los Alamos National Laboratory
Los Alamos, NM 87545, USA

Editors

Library of Congress Cataloging-in-Publication Data. Khriplovich, I. B. (Iosif Bents͡ionovich) *CP* violation without strangeness : electric dipole moments of particles, atoms, and molecules / Iosif B. Khriplovich, Steve K. Lamoreaux. p. cm. – (Texts and monographs in physics, ISSN 0172-5998) Includes bibliographical references and index. 1. *CP* violation (Nuclear physics) 2. Time reversal. 3. Dipole moments. I. Lamoreaux, Steve Keith. II. Title. III. Series. QC793.3.V5K47 1997 539.7'25–dc21 97-28480

ISSN 0172-5998
ISBN-13: 978-3-642-64577-8 e-ISBN-13: 978-3-642-60838-4
DOI: 10.1007/978-3-642-60838-4

Softcover reprint of the hardcover 1st edition 1997

Typesetting: Data conversion by Satztechnik Katharina Steingraeber, Heidelberg
Cover design: *design & production* GmbH, Heidelberg
SPIN: 10520214 55/3144-5 4 3 2 1 0 - Printed on acid-free paper

Preface

Electric dipole moments (EDMs) have interested physicists since 1950, when it was first suggested that there was no experimental evidence that nuclear forces are symmetric under parity (P) transformation. This question was regarded as speculative because the existence of an EDM, in addition to P violation, requires a violation of time-reversal (T) symmetry.

In 1964 it was discovered that the invariance under CP transformation, which combines charge conjugation (C) with parity, is violated in K-meson decays. This provided a new incentive for EDM searches. Since the combined operations of CPT are expected to leave a system invariant, breakdown of CP invariance should be accompanied by a violation of time-reversal symmetry. Thus there is a reason to expect that EDMs should exist at some level. The original neutron EDM experiments were later supplemented with checks of T invariance in atoms and molecules. These investigations are pursued now by many groups. Over the years, the upper limit on the neutron EDM has been improved by seven orders of magnitude, and the upper limit on the electron EDM obtained in atomic experiments is even more strict.

Even without the discovery of the effects sought, the neutron and atomic experiments have ruled out most models of CP violation suggested to explain the effects in K-meson decays; in fact, one could argue that the neutron EDM measurement has ruled out more theoretical models than any other experiment in the history of physics. As to the mechanism of CP violation incorporated in the standard model of electroweak interactions, which is most popular at present, the prediction for the neutron EDM is six orders of magnitude below the present experimental bound. The gap for the electron EDM in this model is much larger.

But does this mean that the experiments discussed in this book are of no serious interest for the elementary particle physics, that they are nothing but mere exercises in precision spectroscopy? Just the opposite. It means that these experiments now, at the present level of accuracy, are extremely sensitive to possible new physics beyond the standard model, physics to which the kaon decays are insensitive. These experiments have been described as "the poor man's high-energy physics" [1], insofar as they are typically done by relatively small collaborations (less than 10 participants), with a relatively small budget. They certainly provide valuable information complementary to the more traditional high-energy experiments.

It is essential that there are no fundamental restrictions to the accuracy that can be obtained in a precision spectroscopy experiment, barring counting statistics and systematic effects. Of course, at any given time, any specific experiment runs into technical limitations, but over the last fifty years for example, the limit on the neutron EDM has been reduced by nearly two orders of magnitude per decade. It is interesting to note that, at each stage of the progress, the technical difficulties were enormous and certainly the maximum sensitivity, given the state of the art, was achieved. Nonetheless, after each stage new ideas would arise, and experimentalists had the courage and support to put those ideas into practice; taken together with the great theoretical progress in regard to interpretation of the experiments, the study of EDMs has been, and will remain, a fruitful and exciting subject.

We believe that this circumstance is by itself a good reason to write a book on the subject. Besides, we hope that the wide-ranging collection of physical problems presented here will be useful for advanced studies of various branches of physics.

It is impossible to list here all those who generously shared with us their knowledge and understanding of the subject. However, we cannot help mentioning some persons whose collaboration has influenced so much the content of this book: R.S. Conti, V.V. Flambaum, E.N. Fortson, R. Golub, B.R. Heckel, V.M. Khatsymovsky, J.M. Pendlebury, M.E. Pospelov, N.F. Ramsey, O.P. Sushkov, A.I. Vainshtein, A.S. Yelkhovsky, and A.R. Zhitnitsky. We owe special thanks to D.P. DeMille for critical comments on the manuscript. To our friends, those whose names are given here and those who we could not mention, we owe our deep and sincere gratitude.

Novosibirsk, Russia *I.B. Khriplovich*
Los Alamos, New Mexico, USA *S.K. Lamoreaux*
May 1997

Contents

Notation

> Physics is essentially a simple science.
> The main problem in it is to understand
> which symbol means what.
>
> *V.A. Fock*

Greek subscripts are four dimensional. They run through values $0, 1, 2, 3$. Latin subscripts are three dimensional.

All four-vectors are considered to be contravariant.

The scalar product is: $a_\mu b_\mu = a_0 b_0 - a_m b_m$;

$\epsilon_{\mu\nu\kappa\lambda}$ is a completely antisymmetric fourth-rank tensor; $\epsilon_{0123} = 1$.

The four-dimensional gradient is:

$$\partial_\mu = (\partial_0,\ -\partial_m) = \left(\frac{\partial}{\partial t},\ -\frac{\partial}{\partial x_m}\right).$$

σ_m are the Pauli matrices.

The Dirac matrices are defined in the following way:

$$\gamma_0 = \begin{pmatrix} I & 0 \\ 0 & -I \end{pmatrix},\ \gamma_0^2 = I;\ \gamma_m = \begin{pmatrix} 0 & \sigma_m \\ -\sigma_m & 0 \end{pmatrix},\ \gamma_m^2 = -I;$$

$$\gamma_5 = -\mathrm{i}\gamma_0\gamma_1\gamma_2\gamma_3 = \begin{pmatrix} 0 & -I \\ -I & 0 \end{pmatrix},\ \gamma_5^2 = I;$$

$$\sigma_{\mu\nu} = \frac{1}{2}(\gamma_\mu\gamma_\nu - \gamma_\nu\gamma_\mu);\ \alpha_m = \gamma_0\gamma_m = \begin{pmatrix} 0 & \sigma_m \\ \sigma_m & 0 \end{pmatrix}.$$

In Chaps. 2, 5, 8, 10, 11 and Sects. 1.4, 9.3, 9.5, we use the system of units with $\hbar = 1,\ \ c = 1$.

The nuclear unit of length (Fermi) $1\ \text{fm} = 10^{-13}$ cm.

m_p is the proton mass. The corresponding Compton wavelength is

$$1/m_\mathrm{p} = 2.103 \times 10^{-14}\ \text{cm} = 0.2103\ \text{fm}.$$

m is the electron mass.

e is the (modulus of) electron charge, $\alpha = e^2 = 1/137.036$.

$G = 1.027 \times 10^{-5}/m_\mathrm{p}^2$ is the Fermi weak interaction constant.

$a = 1/(m\alpha) = 0.529 \times 10^{-8}$ cm is the Bohr radius.

The characteristic atomic energy, the Rydberg, is

$$\mathrm{Ry} = \frac{m\alpha^2}{2} = 13.605\ \text{eV} = 109\ 737\ \text{cm} = 3.28984 \times 10^9\ \text{MHz}.$$

The molecule electric dipole moment length scale, the Debye, is

$$D = 0.208 \times 10^{-8}\ e\ \text{cm}.$$

1. Introduction

Yes, I have caught the tiger.
O tiger, do not bray!
And what immortal hand or eye
Could frame his fearful symmetry
I should not like to say.

A.E. Housman

1.1 Overview of CP Violation Without Strangeness

Over thirty years have elapsed since the discovery by Christensen, Cronin, Fitch and Turlay of violation of the combined symmetry of charge conjugation (C) and parity (P) in the decay of the K^0 meson [2], and the exact origin of this symmetry violation remains an enigma, although the so-called standard model of electroweak interactions can describe these processes phenomenologically. Let us mention at once that the CP violation observed in K^0 decays implies a time-reversal (T) asymmetry, for, as we will discuss later, there are strong reasons to believe that the combined symmetry of CPT should not be violated.

Whether the standard model is the final description of our Universe remains an open question. Tests of the standard model in regard to its successful description of CP violation can be broadly classified into two categories. The first are based on precise measurements of the decay modes of the K^0 and other particles with nonvanishing values of such quantum numbers as strangeness, charm, beauty or truth, i.e. containing s, c, b, or t quarks. Such experimental studies occur in the arena of high energy physics and make use of the associated large-scale accelerators and detectors.

The second class of experiments are those based on precision measurements of the properties of "ordinary" matter, that is, matter for which the expectation values of the strangeness, charm, beauty, and truth quantum numbers for each of the constituents is zero – hence the rather whimsical title of this book. [1] Experiments in this second class tend to be rather small

[1] Therefore, to be precise, the title of the book should be supplemented with the words "without charm, without beauty, without truth". However, the subject certainly has charm at least for us, we hope to communicate its beauty to the readers, and believe that there is truth in it.

scale, sometimes described as "table-top", and perhaps the most exotic system studied here is the neutron, available copiously if not parasitically at several research reactors around the world.

On the other hand, it has been proposed (originally by Sakharov [3]) that the matter–antimatter asymmetry observed in our Universe could have arisen from a CP violating interaction, together with a baryon number non-conserving process, active at an early stage of the Big Bang. Given that these arguments are intricately entwined with cosmology, and that there is some dispute as to whether the CP violation (or possibly violations) implied here is the same as that responsible for the effects observed in K^0 decays, we will not consider matter–antimatter asymmetry, restricting our attention to earth-based experiments.

Another question we will not discuss is the possible connection between the microscopic T violation, as observed in K^0 decay, and irreversibility in thermodynamics and quantum mechanical measurements. Although one might expect that the microscopic T violation plays some role in establishing the observed "arrow of time" associated with our Universe, how to make this connection is at present completely unknown.

1.2 The Neutron Electric Dipole Moment: Early History

The history of the search for the electric dipole moments (EDM) can be traced back to 1950. That year Purcell and Ramsey [4] asked the seminal question: What is the experimental evidence for the absence of nuclear EDMs? In fact, at that time a theoretical argument in favor of its absence based on parity conservation was well known. It is as follows. In quantum mechanics, a non-degenerate system is characterized in the rest frame by the internal angular momentum $\boldsymbol{j}$, the only vector in the system; if we could specify a direction in addition to and independent of the angular momentum, the system would be degenerate because additional quantum numbers would be required to specify a state, contrary to our assumption. Therefore, the EDM $\boldsymbol{d}$ can be directed only along $\boldsymbol{j}$ (or opposite to it):

$$\boldsymbol{d} = \beta \boldsymbol{j}. \tag{1.1}$$

To be more precise, when a stationary quantum-mechanical state is characterized by the usual quantum numbers $\boldsymbol{j}^2 = j(j+1)$ and j_z, the EDM projection can be only

$$d_z = \beta j_z. \tag{1.2}$$

However, under parity transformation (inversion of the space coordinates) a polar vector $\boldsymbol{d}$ changes sign, while an axial vector $\boldsymbol{j}$ does not. This means that $\boldsymbol{d}$ should vanish. So, the nuclear EDM problem was directly related then to that of parity conservation, which was considered at that time by most physicists as being beyond any doubt. The paper [4] was the first to describe the lack of experimental evidence for this symmetry, and the question was to some degree inspired by the Dirac magnetic monopole.

In fact, somewhat earlier, in 1949, Dirac himself questioned seriously the validity of P and T symmetries. He wrote "I do not believe there is any need for physical laws to be invariant under reflections in space and time although the exact laws of nature so far known do have this invariance" [5]. And his own interest in the problem stemmed also from the monopole.

The authors of [4] not only posed the question, but started an experimental search for the neutron EDM d_n which resulted in an upper limit of [6]

$$d_n/e < 5 \times 10^{-20} \text{ cm}.$$

The true fall of discrete symmetries started in 1956 with the famous $\theta - \tau$ paradox in the decays of K mesons. The masses of the so-called θ particle decaying into two π mesons, and τ particle decaying into three pions, turned out to be suspiciously close. It would be tempting to announce that θ and τ are the same particles, but the final states, 2 π and 3 π, had opposite parities. The decisive step was made by Lee and Yang [7] who attributed the paradox to a P violation in the weak interactions responsible for the decays. Immediately, Ramsey suggested an experimental search for parity nonconservation in the β decay of polarized ^{60}Co (the story is described in [8]). If parity were not conserved, the distribution of β electrons could depend on $\boldsymbol{j} \cdot \boldsymbol{p}_e$, where $\boldsymbol{j}$ is the nuclear spin direction and $\boldsymbol{p}_e$ is the electron momentum. In fact, this proposal was well received; Feynman gave Ramsey 50-to-1 odds that parity violation would not be experimentally observable. Perhaps inspired by such certainly lucrative odds, Ramsey and collaborators began to set up an experiment at Oak Ridge, but this experiment was regarded as speculative and almost immediately discontinued to allow an investigation of some new effects observed in fission. Within a few months parity nonconservation was directly demonstrated by three nearly simultaneous experiments: by Wu et al. [9] in the β decay of ^{60}Co, by Garwin, Lederman and Weinrich [10], and by Friedman and Telegdi [11] in the π and μ decays.

Immediately after the observation of parity violation, it was shown by Landau that EDMs are forbidden by time-reversal symmetry [12]. The time-reversal asymmetry implied by the existence of an EDM is easily seen in equations (1.1) and (1.2), where the left- and right-hand sides behave differently under time reversal; $\boldsymbol{d}$ stays intact (just as a radius-vector), and $\boldsymbol{j}$ changes sign. However, it was pointed out again [13, 14] that T invariance should also be checked experimentally, as had been done in the case of parity.

As we have already mentioned, in 1964 CP violation itself was observed in K^0 decay, and with this, EDMs and other possible time-reversal manifestations became of renewed interest, if not acquiring a certain legitimacy.

1.3 Molecular Electric Dipole Moments and CP Violation

Here perhaps it is time for us to dwell on the frequently posed question: How can one claim that a quantum-mechanical system cannot possess a dipole moment when values of this quantity for polar molecules are presented in numerous handbooks? The answer is that the common visual image of the polar molecule with a dipole moment refers to the case when the external field is sufficiently strong to orient this molecule. This is not a pure eigenstate of the Hamiltonian for the isolated molecule, but a superposition of such eigenstates. The EDM of interest to us is that defined in the weak-field limit, i.e., that referring to a pure state.

What happens in this limit is most easily explained in the case of a diatomic polar molecule. Here obviously the only possible orientation of the dipole moment is along the molecular axis. On the other hand, the rotational angular momentum of a diatomic molecule is directed perpendicular to the axis. Therefore, in a stationary state, with a fixed rotational angular momentum, the EDM from the handbooks averages to zero.

More subtle is the situation with polyatomic molecules. Let us start with a symmetrical top molecule; we take ammonia (NH_3) as an example. Being treated as a rigid body, it can be characterized by a specific quantum number k, the projection of the rotational angular momentum onto the symmetry axis (see, e.g., [15], §103). Obviously, k is a T-odd pseudoscalar. Therefore, the relations (1.1) and (1.2) become possible, together with the existence of an EDM, with β being proportional to k. However, for a symmetrical top, the energy levels with $k \neq 0$ are degenerate with respect to the sign of k. In real life this degeneracy is lifted, for instance by the N atom tunneling through the H_3 plane. The resulting nondegenerate states are

$$\frac{1}{\sqrt{2}}\left(|k\rangle \pm |-k\rangle\right). \tag{1.3}$$

In these states, where $|k\rangle$ and $|-k\rangle$ are present with equal weights, the EDM indeed becomes zero, as it should be in any nondegenerate quantum mechanical state. However, when the energy of interaction with an external electric field exceeds the splitting between symmetrical and antisymmetrical states (1.3), the Stark effect becomes linear, i.e. the molecule acquires an EDM.

Finally, in the completely general case of (asymmetrical top) polyatomic molecules, no projection of the rotational angular momentum onto any principal axis is conserved. Corresponding to any given rotational angular momentum J, there are $2J+1$ nondegenerate stationary states. Although there is no conserved quantum number k, all these states still fall into two independent families (see again [15], §103). In one family, the wave functions are even in k, i.e., they can be constructed as linear combinations of

$$\frac{1}{\sqrt{2}}\left(|k\rangle + |-k\rangle\right)$$

with different k. The wave functions of the other family, being odd under $k \leftrightarrow -k$, are constructed from

$$\frac{1}{\sqrt{2}}\left(|k\rangle - |-k\rangle\right).$$

Obviously, the EDM expectation value vanishes in any of the stationary states where $|k\rangle$ and $|-k\rangle$ are present with equal weights.

In the above discussion we relied explicitly on arguments well known from standard molecular physics. However, it is worth mentioning that those arguments in fact are based essentially on the assumptions of P and T invariance.

After convincing (hopefully) the reader that the subject of this book does make sense, we could come back to the subject itself. But first one more topic is worthy of discussion.

1.4 T-Odd Effects Without CP Violation

The title of this section may seem puzzling. However, in classical physics we often come across phenomena noninvariant under T reversal without any CP violation. It is sufficient to recall Ohm's law (to say nothing of, for example, friction):

$$\boldsymbol{j} = \sigma \boldsymbol{E}. \tag{1.4}$$

This equation is obviously T violating since under T reversal $\boldsymbol{j}$ changes sign, but $\boldsymbol{E}$ does not. The explanation is well known: We are dealing here with the irreversible increase of entropy. It is only due to dissipation that an applied electric field induces a constant current. In other words, the current transmission is in fact a multichannel reaction; internal degrees of freedom are excited in the conductor, which results finally in its heating.

Another multichannel effect, the interference between amplitudes of different angular momenta, induces what are called T-odd correlations due to final-state interactions.

All these questions constitute the subject of the detailed book [16]. Here we will address only those problems that are somehow related to the study of dipole moments and which admit simple, intuitive considerations.

1.4.1 Can an Unstable Particle Have a Dipole Moment?

This question was raised by Zel'dovich many years ago [17]. He pointed out that the usual T invariance argument which forbids the interaction

$$H_{\mathrm{d}} = -d\,\langle\boldsymbol{\sigma}\rangle\boldsymbol{E},$$

is not applicable to an unstable particle. Indeed, such a particle has a wave function that decays exponentially in time, and is surrounded by an outgoing wave of its decay products. Under T reversal, it transforms not into the same particle with the opposite spin direction, but into something else, into a state with an exponentially increasing wave function, which is surrounded by incoming decay products. Therefore, if P invariance, which also forbids EDM, is violated, one may expect that an unstable particle has nonvanishing dipole moment.

This conclusion was criticized and eventually discarded [18–22]. Still, we believe that the problem is sufficiently instructive to be discussed in the present book. However, as distinct from the rather sophisticated treatment of the above-mentioned works, we will resort to what, in our opinion, are simpler though less rigorous arguments.

Let us consider the well known example of a parity doublet mixed by a P-odd interaction: the $2s_{1/2}$, $2p_{1/2}$ levels of atomic hydrogen (see, for instance, Sects. 2.2 and 2.3 of [23]). An external electric field directed along, say, the z axis, mixes the $2s_{1/2}$, $2p_{1/2}$ states with the same projection j_z of the total angular momentum. Under the common phase definition, the corresponding sp matrix element, V, is real (see Sect. 8.3.1 below). These states are also mixed by a P-odd, T-even weak interaction. Under the same phase convention, the sp matrix element of the interaction, $\mathrm{i}W$, is imaginary.

Thus, the Hamiltonian matrix for this two-level system can be written as

$$\begin{pmatrix} E_s - \mathrm{i}\,\Gamma_s/2 & V + \mathrm{i}\,W \\ V - \mathrm{i}\,W & E_p - \mathrm{i}\,\Gamma_p/2 \end{pmatrix}. \tag{1.5}$$

The crucial point is that, although the metastable $2s_{1/2}$ and $2p_{1/2}$ levels have finite widths $\Gamma_{s,p}$, and their energies are complex, the interaction (off-diagonal) part of the Hamiltonian matrix (1.5) is hermitian. And this is sufficient to guarantee that the eigenvalues of this matrix depend on $V^2 + W^2$ only. There is no energy shift linear in the interaction V with the electric field, no linear Stark effect. In other words, in spite of being unstable, the system has no dipole moment.

This does not mean, however, that the nonvanishing level widths do not bring qualitatively new features into P-odd effects in this system. For instance, the parity-violating admixture to the $2s_{1/2}$ state is proportional to

$$\frac{\mathrm{i}\,W}{E_s - E_p - \mathrm{i}\,(\Gamma_s - \Gamma_p)/2} = \frac{\mathrm{i}\,W\,\Delta E - W\,\Delta\Gamma}{(\Delta E)^2 + (\Delta\Gamma)^2}. \tag{1.6}$$

The second term in the numerator of this expression, proportional to $\Delta\Gamma$, is as P-odd as the first one. However, being shifted in phase by $\pi/2$, it behaves differently under T reversal: it is T-odd. In this way P-odd, T-odd correlations with an external electric field do indeed arise in radiative transitions, being proportional to the widths, or to the final state interactions. Their appearance is as natural as, for instance, that of the well known P-even correlation of spin with the normal to the reaction plane in scattering processes beyond the

Born approximation. Detailed theoretical studies of these effects have been carried out for hydrogen and hydrogen-like ions [24–25b]. Unfortunately, the effects are too small (as are P-odd, T-even effects in hydrogen) to be observed experimentally.

This problem is related amusingly to another one, widely discussed in recent years. We refer to the possibility of experimental tests of T invariance in neutron elastic scattering from polarized nuclei. It was pointed out in particular that the P-odd, T-odd correlation

$$\boldsymbol{\sigma} \cdot \boldsymbol{I} \times \boldsymbol{p} \tag{1.7}$$

in the forward scattering cannot be mimicked by the final state interaction [26]. (In the expression (1.7), $\boldsymbol{\sigma}$ is the neutron spin, $\boldsymbol{p}$ is its momentum, and $\boldsymbol{I}$ is the spin of the target nuclei.) At first sight, this conclusion may seem puzzling. Indeed, T-odd correlations caused by the mixing (1.6) should arise in the decay of the compound nucleus which is the intermediate state of the reaction discussed.

Let us recall however that this decay is only the second stage of the scattering reaction. During this stage the wave function of the compound nucleus decreases with time, which is described by the finite width Γ of the intermediate state. But the scattering process also has the first phase during which the intermediate state is created, so that its wave function increases with time. This growth in time can be formally described by a width of opposite sign. Then, the T-odd correlations due to mixing (1.6) during the first and second stages of the scattering are thus of opposite sign and they just cancel. This conclusion was actually reached long ago [19] (see also [20]) by detailed investigation in the framework of the Lee model.

It is worth mentioning that the above argument in no way contradicts the existence of the well known (P-even) correlation

$$\boldsymbol{\sigma} \cdot \boldsymbol{p}' \times \boldsymbol{p}.$$

Here the initial $\boldsymbol{p}$ and final $\boldsymbol{p}'$ momenta are noncollinear, so that the effects at the initial and final stages of the scattering do not cancel.

1.4.2 Spurious EDM Effect Due to Nuclear Anapole Moment

Ohm's law can lead to an imitation of the EDM interaction with an external electric field $\boldsymbol{E}$ [27]. Indeed, when this field is applied to a material, by Ohm's law, a leakage current $\boldsymbol{j}$, parallel to $\boldsymbol{E}$, is created; this in turn interacts, for instance, with a nuclear spin $\boldsymbol{I}$. This P-odd, but T-even interaction is induced mainly by contact electromagnetic interactions of electrons with the anapole moment of a nucleus, which is a P-odd electromagnetic moment induced by parity-violating nuclear forces (see [23], Chap. 8). The corresponding Hamiltonian can be written

$$H_{\mathrm{a}} = \frac{G\kappa_{\mathrm{a}}}{\sqrt{2}} \tilde{K} \boldsymbol{n} \cdot \boldsymbol{v}\, \delta(\boldsymbol{r}). \tag{1.8}$$

Here $\kappa_a \approx 0.3$ is a dimensionless constant characterizing the nuclear anapole moment, $\tilde{K}$ is a Clebsch–Gordan factor close to unity, $\boldsymbol{n}$ is the unit vector of nuclear polarization, and $\boldsymbol{v}$ is the electron velocity operator.

It is clear now that a well-defined electric current $\boldsymbol{j}$ will induce a frequency shift between the Zeeman levels of a nucleus, with quantization axis along the current flow. If one then does an NMR experiment with an applied magnetic field along the current, there will be an anomalous EDM-like interaction with $\boldsymbol{j}$ parallel to $\boldsymbol{E}$. A P violation will interfere with a T-odd effect of macroscopic origin which is due to irreversible scattering of electrons in the conductor.

The corresponding effective shift of the NMR frequency, as created by a macroscopic current I flowing through the cross-section S, can be estimated as

$$\Delta f = \frac{G\kappa_a Z^2 R}{\sqrt{2}} \frac{I}{eS}. \tag{1.9}$$

Besides obvious dimensional arguments, we have taken into account in this formula the $Z^2 R$ enhancement of the electron current density near the nucleus; the relativistic factor R in it becomes essential at high Z and reaches values about 10 at $Z \approx 80$. Numerically, for ^{199}Hg the shift is of magnitude

$$\Delta f = 4 \times 10^{-10} \frac{I}{S} \text{ cm}^2 \text{ Hz A}^{-1}. \tag{1.10}$$

So, a current of 1 A flowing in a cylinder of 1 cm diameter, shifts the frequency by $\Delta f = 4 \times 10^{-10}$ Hz, which corresponds to an apparent nuclear EDM of $10^{-28} e$ cm.

This spurious effect is much too small to concern us. On the other hand, it is tempting to speculate that, in principle, this effect could be used for the detection and measurement of nuclear anapole moments.

2. Kinematics of Discrete Symmetries

Stop talking!
Write down the Hamiltonian.
Students' folklore

2.1 *CPT* Theorem: Intuitive Approach

At present, the breakdown of P, C, and even CP invariance is firmly established. But what has survived up to now is the invariance under combined CPT. There are no serious experimental reasons to doubt this invariance and, moreover, there are very strong theoretical reasons, based on sound physical grounds, to expect it. The statement of this invariance is known as the CPT theorem [28–30].

Instead of a formal proof of this theorem (which is beyond the scope of this book), let us present intuitive, physical arguments in its favor.

Our space-time is four dimensional, and there is a well known geometrical fact that for an even dimensional space, reflection of all axes is equivalent to a rotation. This can be easily demonstrated for a plane, i.e., a two dimensional space: Under total reflection, both coordinate axes change sign, and exactly the same happens for a rotation by π around the origin.

It would therefore be tempting to assume that PT reflection is equivalent to a rotation in our four dimensional space. In particular, for the rotation in question, all components of any 4-vector should change signs. But what happens under PT with the usual vector current j_μ? Let us recall that the charge density ρ does not change under either P, or T, while the three dimensional current $\boldsymbol{j}$ changes sign under either operation. Thus,

$$j_\mu = (\rho, \boldsymbol{j}) \xrightarrow{P} (\rho, -\boldsymbol{j}) \xrightarrow{T} (\rho, \boldsymbol{j}),$$

and obviously the PT guess does not work. The reason is that our space-time is not so simple: it is *pseudo* euclidean, and the time coordinate is not exactly equivalent to a space coordinate. In particular, while changing the sign of a three dimensional momentum is nothing special, changing the sign of an energy is known to be related to the transition from particle to antiparticle.

But then it looks reasonable to try one more step: C conjugation, which changes the sign of the electromagnetic four-current, and we get immediately

$$j_\mu \xrightarrow{CPT} -j_\mu .$$

So, it appears indeed that in our pseudoeuclidean space-time, it is the CPT operation which is equivalent to the reflection of all four axes.

However, in addition to a usual four-vector changing sign, an axial four-vector should also change sign under four dimensional rotation, and therefore under reflection of all four axes. To check this, let us recall that, say, for a Dirac particle, the axial current is simply the covariant four dimensional spin vector. Therefore, its three dimensional components behave as the usual nonrelativistic spin vector $\boldsymbol{s}$: they change sign under T, but not under P. Its time component, being just the Lorentz-transformed nonrelativistic spin, is proportional to $\boldsymbol{s} \cdot \boldsymbol{v}$ and therefore changes sign under P, but does not under T. And if we recall finally the V–A structure of the usual weak interaction which conserves CP only, but not C (or P), we see that the axial current does not change sign under C, contrary to the vector case (a detailed rigorous treatment of C conjugation, as well as P and T operations can be found, e.g., in [31], Sects. 26 and 27). Thus, an axial four-vector transforms under CPT properly:

$$a_\mu = (a_0, \boldsymbol{a}) \xrightarrow{C} (a_0, \boldsymbol{a}) \xrightarrow{P} (-a_0, \boldsymbol{a}) \xrightarrow{T} (-a_0, -\boldsymbol{a}).$$

From now on, after demonstrating the validity of the CPT theorem, we will consider CP invariance and T invariance as synonyms.

Although C is not an exact symmetry of Nature, and in this sense there is no symmetry between particles and antiparticles, still CPT invariance guarantees that particles and antiparticles have equal masses, equal *total* lifetimes, and equal (in modulus) electromagnetic moments. The proof is simple. For instance, a particle mass is nothing more than the expectation value of its total Hamiltonian H in a state at rest with a given spin projection s_z:

$$m = \langle s_z | H | s_z \rangle .$$

Under CPT the Hamiltonian stays intact, and particle states transform into antiparticle states:

$$|s_z\rangle \longrightarrow \langle\overline{-s_z}|, \quad \langle s_z| \longrightarrow |\overline{-s_z}\rangle$$

(T reversal implies the interchange of initial and final states). Due to the rotational invariance, the mass cannot depend on the spin projection, and on its sign in particular. Therefore, we indeed have

$$m = \overline{m}.$$

An analogous line of reasoning applies to the total decay rate, which is proportional to

$$\sum_n \langle s_z|H_\mathrm{d}|n\rangle\langle n|H_\mathrm{d}|s_z\rangle,$$

where H_d is the decay Hamiltonian. An essential feature of this proof is that it refers to the *total* decay rate, i.e. the sum in this expression includes all possible final states $|n\rangle$.

Furthermore, and lastly, the same arguments demonstrate the equality of electromagnetic moments of a particle and its antiparticle.

The rigorous proof of the CPT theorem is based on such general assumptions as Lorentz-invariance and locality. It is therefore no surprise that nobody has succeeded in the construction of a consistent theoretical model of CPT violation. The situation with P and CP nonconservation is quite different: it was never a problem to incorporate them into a theoretical model.

Certainly, CPT invariance should be (and is being) checked experimentally, but this should be done with a recognition of the firm basis of the CPT theorem, and of the extremely serious implications of a violation of this invariance.

2.2 T-Even and T-Odd Electromagnetic Multipole Moments

In this section we will discuss the most general structure of the electromagnetic current, with emphasis on the properties of its various terms under P, T, and C operations. Consider a matrix element J_μ of the electromagnetic current operator between states with a given spin I and momenta k and k'. It can be conveniently expanded in the four independent Lorentz vectors of the problem:

$$p_\mu = (k'+k)_\mu,\ q_\mu = (k'-k)_\mu,\ s_\mu,\ r_\mu = \mathrm{i}\epsilon_{\mu\nu\kappa\lambda}p_\nu q_\kappa s_\lambda. \tag{2.1}$$

Here s_μ is the four dimensional spin operator defined, e.g., for the state with momentum k. It is obtained through the Lorentz transformation of the spin vector $(0, \boldsymbol{I})$ in the rest frame of a particle and can be written as

$$s_\mu = (s_0, \boldsymbol{s}),\ s_0 = \frac{\boldsymbol{I}\cdot\boldsymbol{k}}{m},\ \boldsymbol{s} = \boldsymbol{I} + \frac{(\boldsymbol{I}\cdot\boldsymbol{k})\boldsymbol{k}}{m(E+m)}.$$

Being defined in this way, the vector s_μ obviously satisfies the condition $k_\mu s_\mu = 0$ (in which case, however, $k'_\mu s_\mu \neq 0$). Allowing for the current conservation law which, in the momentum representation, is $q_\mu J_\mu = 0$, the matrix element of the current can be written as follows (see, e.g., [23], Chap. 8):

$$J_\mu = \langle \boldsymbol{I}, \boldsymbol{k}'|p_\mu F_1 + r_\mu F_2 + [q^2 s_\mu - q_\mu(qs)]F_3|\boldsymbol{I}, \boldsymbol{k}\rangle. \tag{2.2}$$

The invariant functions F_i depend on the scalars that can be constructed from the vectors (2.1). Since pq, rp, rq, rs, and sk vanish, there are only two such independent invariants. It is convenient to choose the invariant momentum transfer $t = q^2$ and the hermitian operator $\tau = iq_\mu s_\mu$ as the arguments of F_i. The hermiticity of this operator is most easily seen in the coordinate representation, where $i\psi^\dagger q_\mu \psi \to \partial_\mu(\psi^\dagger\psi)$. The substitution of the operator function $F_i(t,\tau)$ into the expression (2.2) should be accompanied by symmetrization in the noncommuting operators s_μ and τ. In the expansion of $F_i(t,\tau)$ in powers of τ

$$F_i(t,\tau) = \sum_{n=0}^{N_i} f_{i,n}(t)\tau^n \tag{2.3}$$

the highest power N_i is evidently fixed by the spin I:

$$N_1 = 2I, \quad N_2 = N_3 = 2I - 1.$$

Therefore, the total number of electromagnetic form factors $f_{i,n}$ is $6I + 1$.

Let us look now at the behavior of different terms of the current (2.2) under the discrete symmetry operations. It can be easily seen that all three vector structures in formula (2.2) by themselves behave under time reversal identically, just as a current should. On the other hand, only two of these vectors, p_μ and r_μ have correct, current-like, properties under space inversion. As to the operator argument τ of the functions F_i, it changes sign under time reversal since $i\boldsymbol{q}$, or $\boldsymbol{\nabla}$, does not, while $\boldsymbol{s}$, as an angular momentum should, changes to $-\boldsymbol{s}$. Then, τ is a pseudoscalar since q and s are polar and axial vectors, respectively. In virtue of the CPT theorem, it means that τ is a C-even operator. Therefore, all structures in the matrix element (2.2) generated by $F_{1,2}$ are odd under C conjugation, which is natural for the electromagnetic current.

However, the axial vector $q^2 s_\mu - q_\mu(qs)$ behaves properly under T reversal and is therefore C violating. Even terms in the expansion (2.3) for F_3 are P-odd, but T-even; odd terms are P-even, but T-odd. In particular, the term with $n = 0$ corresponds to the so-called anapole moment, characteristic of systems without a center of inversion. Nuclear anapole moments are of great physical interest (see detailed discussion in [23]). Even though anapole moments have yet to be definitively observed, present experimental upper limits on their magnitude [32–35] give valuable information about P-odd nuclear forces [36]. The P-even, T-odd term with $n = 1$ in the F_3 expansion is discussed in [37, 38].

A peculiar property of the C violating multipoles, both T-even and T-odd, is that the interaction generated by them is local [39]. Indeed, electromagnetic interaction of J_μ with the test current j_μ is written as

$$-j_\mu \frac{1}{q^2} J_\mu. \tag{2.4}$$

Since the test current j_μ is conserved, $q_\mu s_\mu = 0$, the term $-q_\mu(qs)$ in J_μ is not operative, while the factor q^2 in $q^2 s_\mu$ cancels the propagator $1/q^2$ in (2.4). The remaining C-odd interaction $-j_\mu\langle I,k'|s_\mu F_3|I,k\rangle$ is a contact one.

The locality of C violating electromagnetic multipoles appears quite natural: a real photon is C-odd and therefore cannot be emitted by a C-even vertex.

Finally, the C violating form factors constitute the only family of electromagnetic multipoles possible for truly neutral particles [40]. Indeed, a neutral particle and its antiparticle have the same C symmetry. Therefore, in the annihilation channel, (i.e., when considering the annihilation of a particle and its antiparticle into a virtual photon, instead of a virtual photon emission by a particle) the C parity of a particle–antiparticle state is even while that of a photon is odd, and the electromagnetic vertex of a truly neutral particle must violate C invariance.

Let us now come back to the C conserving terms in the electromagnetic current. All T invariant electromagnetic multipoles of this type are generated by terms with even n in the $F_{1,2}$ expansions. Therefore, the usual well known P and T-even form factors, such as the charge, magnetic dipole, and electric quadrupole, belong to the families $f_{1,2n}$, $f_{2,2n}$.

In this book, we will be primarily interested in another family of electromagnetic multipoles, those which violate P and T invariance. This family includes the electric dipole and magnetic quadrupole moments. Moments in this family correspond to odd terms of the $F_{1,2}$ expansion.

The total number of invariant form factors $f_{i,n}$, conserving (+) and violating (−) P, T and C invariance can be easily found for a given spin I and are listed in Table 2.1. If two values are given in a line of the table, the first refers to half-integer spin, the second to integer spin. This classification of electromagnetic form factors was first given in [41] by changing to the annihilation channel.

Table 2.1. Possible number of invariant form factors

P	T	C	Number of form factors
+	+	+	$2I+1$
−	+	−	$I+1/2;\ I$
+	−	−	$I-1/2;\ I$
−	−	+	$2I$

In view of our special interest in the EDM, let us present this multipole in a more convenient form than $p_\mu T$. For a spin-1/2 particle the interaction of its EDM d with the electromagnetic field strength $F_{\mu\nu}$ can be written in the following relativistically invariant form:

$$H_{\rm d} = \frac{d}{2}\bar{\psi}\gamma_5\sigma_{\mu\nu}\psi F_{\mu\nu}. \tag{2.5}$$

Its proper normalization and T-odd nature are obvious from the nonrelativistic limit:

$$H_{\mathrm{d}} = -d\langle\boldsymbol{\sigma}\rangle\boldsymbol{E}. \tag{2.6}$$

Indeed, for $t \to -t$

$$\langle\boldsymbol{\sigma}\rangle \to -\langle\boldsymbol{\sigma}\rangle, \quad \boldsymbol{E} \to \boldsymbol{E}.$$

The following derivation of the interaction (2.5) is instructive also. Let us start from the well known expression for the electromagnetic interaction of the anomalous magnetic moment μ:

$$H_{\mu} = \frac{\mu}{2}\,\bar{\psi}\,\mathrm{i}\sigma_{\mu\nu}\psi\,F_{\mu\nu}. \tag{2.7}$$

Also well known is its nonrelativistic limit:

$$H_{\mu} = -\mu\langle\boldsymbol{\sigma}\rangle\boldsymbol{B}. \tag{2.8}$$

To obtain a T-odd interaction (2.6) from this respectable one we have to perform the substitution

$$\boldsymbol{B} \to \boldsymbol{E}.$$

The four dimensional form of this operation is

$$F_{\mu\nu} \to -\tilde{F}_{\mu\nu},$$

where $\tilde{F}_{\mu\nu} = (1/2)\epsilon_{\mu\nu\kappa\lambda}F_{\kappa\lambda}$. Due to the identity

$$\frac{1}{2}\,\epsilon_{\mu\nu\kappa\lambda}\sigma_{\kappa\lambda} = \mathrm{i}\gamma_5\sigma_{\mu\nu},$$

this four dimensional substitution transforms the T-even interaction (2.7) into a T-odd one (2.5).

Finally, a few words on the magnetic quadrupole moment: According to Table 2.1, a MQM can exist only for spin $I \geq 1$. We will define the MQM operator by analogy with the electric quadrupole operator, via the interaction with the corresponding field gradient:

$$\hat{H}_Q = -\frac{1}{6}\hat{Q}_{ij}\nabla_i E_j,$$

$$\hat{H}_M = -\frac{1}{6}\hat{M}_{ij}\nabla_i B_j. \tag{2.9}$$

The tensor $\hat{M}_{ij}$ is expressed as usual via the irreducible second-rank tensor constructed from the spin operator $\hat{I}_i$:

$$\hat{M}_{ij} = \mathcal{M}\frac{3}{2I(2I-1)}\left[\hat{I}_i\hat{I}_j + \hat{I}_j\hat{I}_i - \frac{2}{3}\delta_{ij}I(I+1)\right]. \tag{2.10}$$

The expectation value $\mathcal{M}$ of the operator $\hat{M}_{zz}$ in the state with the maximum spin projection $I_z = I$ will be called magnetic quadrupole moment.

2.3 General Structure of Four-Fermion Operators

Since there is no firmly established theory of CP violation, in this book our approach will be primarily based on its phenomenological description, through the use of effective four-fermion operators. It is appropriate to present here a general classification of these operators. To find the number of scattering amplitudes with different symmetries for fermions of spin 1/2, it is convenient to use the annihilation channel (i.e., to consider the reaction $f_1 + \bar{f}_1 \to f_2 + \bar{f}_2$ instead of the initial $f_1 + f_2 \to f_1 + f_2$) and to classify there the particle-antiparticle states of a given total angular momentum j with respect to P and CP transformations. The P , C and CP parity of the fermion–antifermion system are given by the following relations ([31], §27):

$$P = (-)^{l+1}, \quad C = (-)^{l+s}, \quad CP = (-)^{s+1}, \tag{2.11}$$

where l and s are the orbital and spin angular momenta of the system. In this way the following classification is obtained:

$$\begin{array}{lllll}
(1) & s=0, \; l=j, & P=(-)^{j+1}, & C=(-)^{j}, & CP=- \\
(2) & s=1, \; l=j, & P=(-)^{j+1}, & C=(-)^{j+1}, & CP=+ \\
(3) & s=1, \; l=j+1, & P=(-)^{j}, & C=(-)^{j}, & CP=+ \\
(4) & s=1, \; l=j-1, & P=(-)^{j}, & C=(-)^{j}, & CP=+
\end{array}$$

The CP- and P-even amplitudes, well known to be six in number, are

$$1 \to 1, \quad 2 \to 2, \quad 3 \to 3, \quad 4 \to 4, \quad 3 \to 4, \quad 4 \to 3.$$

In the case of identical fermions this number reduces to five since the amplitudes $3 \to 4$ and $4 \to 3$ are the same.

Furthermore, there are four CP-even, but P-odd amplitudes (two for identical fermions):

$$2 \to 3, \quad 2 \to 4, \quad 3 \to 2, \quad 4 \to 2.$$

However, we will be interested mainly in CP-odd and P-odd amplitudes. There are also four of them (and again two for identical fermions):

$$1 \to 3, \quad 1 \to 4, \quad 3 \to 1, \quad 4 \to 1.$$

Let us present their explicit form which is necessary for our future discussions. One more reason for this explicit presentation is that some confusion concerning not only the structure of these terms, but even their number, exists in the literature.

Three four-fermion operators without derivatives, violating both T and P invariance, are

$$\begin{aligned}
\mathcal{SP} &= \bar{\psi}_1\psi_1\,\bar{\psi}_2 \mathrm{i}\gamma_5\psi_2, \\
\mathcal{PS} &= \bar{\psi}_1 \mathrm{i}\gamma_5\psi_1\,\bar{\psi}_2\psi_2, \\
\mathcal{T} &= \frac{1}{2}\,\epsilon_{\mu\nu\kappa\lambda}\,\bar{\psi}_1\sigma_{\mu\nu}\psi_1\,\bar{\psi}_2\sigma_{\kappa\lambda}\psi_2.
\end{aligned} \tag{2.12}$$

The meaning of the symbols $\mathcal{S}$, $\mathcal{P}$, $\mathcal{T}$ here is obvious: scalar, pseudoscalar and tensor, respectively. Of course, the T and P violating nature of these operators can be directly proven by using the transformation properties of Dirac covariants under the corresponding transformations ([31], §28). We will proceed however somewhat differently. In a nonrelativistic approximation the scalar $\mathcal{S}$ and pseudoscalar $\mathcal{P}$ become

$$\mathcal{S} = \bar{\psi}\psi \to \phi^\dagger\phi, \quad \mathcal{P} = \bar{\psi}\mathrm{i}\gamma_5\psi \to -\frac{1}{2m}\boldsymbol{\nabla}(\phi^\dagger\boldsymbol{\sigma}\phi),$$

where ϕ is the nonrelativistic wave function. Therefore, $\mathcal{S}$ is indeed a scalar since to this approximation it coincides with the number density which does not change under either inversion or time reversal. As to the second covariant, $\mathcal{P}$, it becomes the divergence of the spin density which obviously changes sign under both operations. Therefore, the same is true for the four-fermion operators $\mathcal{SP}$ and $\mathcal{PS}$. The same nonrelativistic reduction demonstrates the hermitian nature of both.

To discuss the transformation properties of the third structure, $\mathcal{T}$, let us recall the magnetic moment interaction (2.7). It is even under inversion and time reversal. Therefore, $i\,\bar{\psi}\sigma_{\mu\nu}\psi$ behaves under both operations like $F_{\mu\nu}$, and the entire operator $\mathcal{T}$ behaves like $F_{\mu\nu}\tilde{F}_{\mu\nu} = 2\boldsymbol{EB}$. The electric field $\boldsymbol{E}$ is a T-even polar vector, while the magnetic field $\boldsymbol{B}$ is a T-odd axial one. Thus, the claimed transformation properties of the third operator $\mathcal{T}$ become obvious.

In addition, one can as easily construct two more T-odd and P-odd operators, but with derivatives. To do this, it is sufficient to substitute a vector current $\bar{\psi}\gamma_\mu\psi$ for the electromagnetic vector-potential A_μ in expression (2.5) for the EDM interaction. The two structures obtained in this way are

$$\begin{aligned}\mathcal{DV} &= \bar{\psi}_1\mathrm{i}\gamma_5\sigma_{\mu\nu}(p_1' - p_1)_\nu\psi_1\bar{\psi}_2\gamma_\mu\psi_2,\\ \mathcal{VD} &= \bar{\psi}_1\gamma_\mu\psi_1\bar{\psi}_2\mathrm{i}\gamma_5\sigma_{\mu\nu}(p_2' - p_2)_\nu\psi_2.\end{aligned} \tag{2.13}$$

Although we have constructed five T-odd, P-odd operators, $\mathcal{SP}$, $\mathcal{PS}$, $\mathcal{T}$, $\mathcal{DV}$, $\mathcal{VD}$, there is no contradiction with the above general argument which demonstrated that there are only four independent structures of this type. The point is that those five operators are interrelated by a linear identity. This identity can be derived by changing to the annihilation channel in the center-of-mass frame and using the two-component reduction of the spinors. It reads [42]

$$s\mathcal{T} + 4(t - u + m_1^2 + m_2^2)(\mathcal{SP} + \mathcal{PS}) + 4(m_1\mathcal{VD} + m_2\mathcal{DV}) = 0. \tag{2.14}$$

Here $s = (p_1' + p_1)^2$ is the center-of-mass energy squared in the annihilation channel, $t = (p_1 - p_2)^2$ and $u = (p_1' - p_2)^2$ are the invariant momentum transfers. The following identity is valid on the mass shell (i.e., for free incoming and outgoing particles):

$$\bar{\psi}\mathrm{i}\gamma_5\sigma_{\mu\nu}(p'-p)_\nu\psi = \bar{\psi}\mathrm{i}\gamma_5(p'+p)_\mu\psi \tag{2.15}$$

which simplifies this calculation, as well as some other calculations to follow.

Finally, with regard to the T-odd and P-even amplitudes, there are obviously only two:

$$1 \to 2, \quad 2 \to 1.$$

To explicitly construct the corresponding operators, it is sufficient to substitute an axial current for the vector current in formulas (2.13). In this way we get the two following four-fermion operators:

$$\begin{aligned}&\bar{\psi}_1\mathrm{i}\gamma_5\sigma_{\mu\nu}(p_1'-p_1)_\nu\psi_1\bar{\psi}_2\gamma_\mu\gamma_5\psi_2,\\&\bar{\psi}_1\gamma_\mu\gamma_5\psi_1\bar{\psi}_2\mathrm{i}\gamma_5\sigma_{\mu\nu}(p_2'-p_2)_\nu\psi_2.\end{aligned} \tag{2.16}$$

Let us note that T-odd, P-even covariant operators necessarily contain derivatives. In other words, their dimension, seven, is higher than the lowest possible dimension, six, of other types of four-fermion operators.

The case of identical fermions ($\psi_1 = \psi_2$) deserves special attention. The T-odd, P-even interaction vanishes if the coefficients of the operators (2.16) are independent of the momentum transfer, in other words, in the local limit. Technically, the direct and exchange amplitudes cancel after the Fierz transformation. The following identities are useful for this proof and for some other calculations to follow:

$$\begin{aligned}&\sigma_{\mu\nu} \times \gamma_\mu(p_2'+p_2)_\nu = \gamma_5(p_1'+p_1)_\mu \times \gamma_\mu\gamma_5,\\&\sigma_{\mu\nu} \times \gamma_\mu\gamma_5(p_2'+p_2)_\nu = -m_2\gamma_5\sigma_{\mu\nu} \times \sigma_{\mu\nu} + \gamma_5\sigma_{\mu\nu}(p_1'-p_1)_\nu \times \gamma_\mu.\end{aligned} \tag{2.17}$$

We omit here the evident field operators ψ; the first and second spin-momentum operators in both expressions refer to the first and second particles, respectively.

A more intuitive way to demonstrate the vanishing of the T-odd, P-even interactions for identical fermions of spin 1/2 in the local limit is as follows. Due to the Pauli principle, the two-particle states here are either singlet with even orbital angular momenta L and correspondingly of even parity, or triplet with odd orbital angular momenta and of odd parity. Being P-even, the interaction therefore does not mix singlet and triplet states. On the other hand, this interaction is a scalar and therefore cannot mix the singlet states among themselves. It mixes only triplet states, ${}^3L_{L+1}$ with ${}^3(L+2)_{L+1}$. Even at the smallest possible orbital angular momentum, $L = 1$, the product of the radial wave functions, ${}^3P_2$ and ${}^3F_2$, behaves as r^4 for $r \to 0$. Since the interaction (2.16) contains only two spatial derivatives (as can be checked by the reduction of the spinors to the two-component form), in the local limit (which corresponds to the presence of $\delta(\boldsymbol{r})$ in the coordinate representation) it cannot mix ${}^3P_2$ and ${}^3F_2$ states, and we conclude that the interaction vanishes for all $L \geq 1$ because of the increasing power of r in the wave function product as $r \to 0$.

If one takes into account the possible dependence of the coefficients of the operators (2.16) on the invariant momentum transfers t and u in the direct and exchange channels, which corresponds to going beyond the local limit, the T-odd, P-even interaction of identical fermions no longer vanishes. However, being proportional (at least) to $t - u$, this interaction is described by an operator of dimension nine or higher.

3. General Features of EDM Experiments

> Most things in life –
> automobiles, mistresses, cancer
> – are only important to those who have them.
> *John Kenneth Galbraith*

In the search for EDMs, the experiments that have been carried out or proposed cover a wide range of systems. Before describing any of them in full detail, a number of general principles which can be applied to any EDM experiment will be outlined; this will aid in making comparisons among experiments and in assessing future possibilities. In this regard, specific features of a number experiments will be introduced; full descriptions of these and other experiments will be given in the appropriate later chapters. A number of reviews [43–45] have been published on the subject and the reader is referred to these also.

3.1 Interaction of an EDM with an Electric Field

In order to detect an EDM, it is necessary to subject the species under consideration (neutron, atom, molecule, etc.) to an electric field and then to look for an appropriate response. In the presence of a non-zero EDM, an electric Zeeman effect occurs in addition to the usual magnetic Zeeman effect, and the Hamiltonian of the system is

$$H = -(\mu \boldsymbol{B} + d\boldsymbol{E}) \cdot \frac{\boldsymbol{F}}{|F|}, \tag{3.1}$$

where $\boldsymbol{B}$ and $\boldsymbol{E}$ are applied static magnetic and electric fields, μ is the magnetic moment, F is the net angular momentum of the species, and d represents an EDM.

The art of all EDM measurements is in the separation of spurious electric field effects from a true EDM effect. The spurious effects can be made quite small; this illustrates an advantage of EDM experiments over T violation study involving β decay or neutron transmission [46] where the sought T violation signal cannot be turned on and off and appears alongside other allowed processes.

In writing (3.1), we have ignored, for example, changes in the species internal structure due to the application of the electric field (electric tensor polarizability). Also ignored is a possible static electric quadrupole moment; these two possible effects indicate some of the advantages of working with spin-1/2 systems where the only possible electromagnetic moment is the magnetic dipole. For a spin-1/2 system, there is no energy shift between $m_F = \pm 1/2$ due to application of an electric field, and therefore no directly observable effect. Also we have assumed that the net species charge is zero; something with a net charge, in any reasonable electric field, would simply be accelerated out of the measurement apparatus.

A typical experimental observable is the change in Larmor precession frequency associated with a reversal of $\boldsymbol{E}$ relative to $\boldsymbol{B}$; this is an energy shift correlated with the quantity $\boldsymbol{E} \cdot \boldsymbol{B}$, a P- and T-odd quantity. An EDM of 1×10^{-26} $e\,\mathrm{cm}$ would produce a relative change in precession frequency, on reversal of $\boldsymbol{E}$ relative to $\boldsymbol{B}$, of 1×10^{-7} Hz when $E = 10$ kV/cm. This frequency shift corresponds to a magnetic field of about 10^{-10} Gauss for a neutron or diamagnetic atom, or about 10^{-13} Gauss for a paramagnetic atom. Given that the Earth's magnetic field is of order 0.5 Gauss, we see immediately that magnetic field control is crucial for any EDM experiment.

Other EDM observables are changes in position or momentum of a neutral species interacting with an electric field gradient; such effects have been sought in neutron scattering experiments. The magnitude of the force $\boldsymbol{f}$ is simply given by the gradient of (3.1), and therefore detection of an EDM force can ultimately be associated with an energy shift.

3.1.1 Sensitivity Limit Due to the Uncertainty Principle

Let us start from very basic principles. Consider a single spin-1/2 species precessing in a magnetic field, as depicted in Fig. 3.1. We can use this picture because a spin-1/2 system is fully described by the vector model; the magnetization (or net spin polarization) in a density matrix formulation is represented by a tensor of rank one, and this is fully represented by a vector. For higher spins, the density matrix can be expanded in higher multipolarities; the rank one component of the expansion is of principal interest in regard to EDM measurements, since (3.1) is of vector character. We can use a simple vector model for the precessing spin because the expectation values from quantum mechanics obey the laws of classical mechanics. (See [47], Complement $\mathrm{E_{IV}}$, for a full discussion of the spin-1/2 density matrix.)

If we let the spin precess for a time T in an applied magnetic field B_0, a net phase $2n\pi + \phi$ accmulates, where n is an integer. We assume the frequency is known well enough to establish n perfectly, and our task is to determine ϕ. Clearly, the final uncertainty in f, the precession frequency, is

$$\delta f = \frac{\delta\phi}{2\pi T}, \tag{3.2}$$

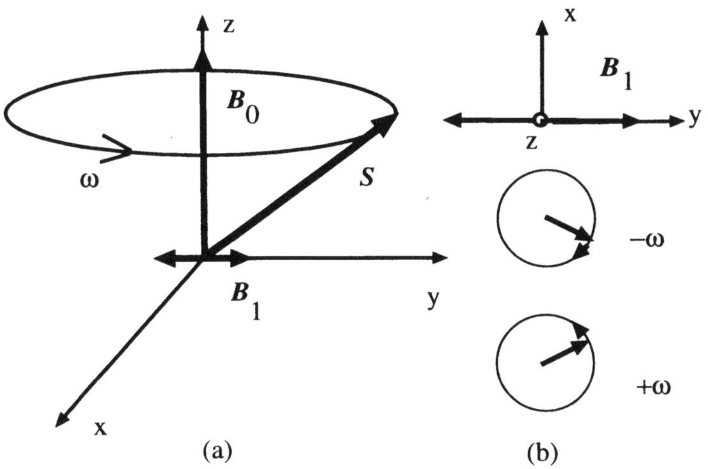

Fig. 3.1a,b. Classical vector model of a spin-1/2 system precessing in a magnetic field. (a) A static field is applied along the z axis, and an oscillating field can be applied along the y axis. (b) The oscillating field can be represented as two counter-rotating fields

where $\delta\phi$ is the uncertainty in ϕ. To determine ϕ, a spin analyzer is introduced along $\hat{y}$, the direction nearly perpendicular to the final spin direction. The probability of finding the spin along $\pm\hat{y}$ is

$$P_{\pm} = (1 \pm \sin\phi)/2 \approx (1 \pm \phi)/2, \tag{3.3}$$

where the approximation applies when the final spin direction is nearly along $\hat{x}$; we will assume hereafter that this is the case. We should remember that the spin analysis is a quantum measurement, and the result of any measurement is $P_+ = 1$, $P_- = 0$ or $P_+ = 0$, $P_- = 1$. If we imagine repeating the precession/measurement sequence many times, on average

$$P_+ - P_- = 2P_+ - 1 = \phi;$$

the uncertainty in ϕ for any measurement is given by twice the uncertainty in P_+; since the average of $P_+ = 1/2$ and its range is 0 to 1, the uncertainty in P_+ is 1/2. Therefore, the uncertainty in ϕ is exactly one, and a single measurement gives a frequency uncertainty of

$$\delta f = \frac{1}{2\pi T}, \tag{3.4}$$

which exactly corresponds to the Heisenberg energy–time uncertainty principle $\delta E \times T = \hbar$. This is a case where the best possible resolution is obtained. Repeating the measurement N times will allow determination of the frequency to $\delta f/\sqrt{N}$. It is interesting to note that the above simple argument can be applied to any quantum energy measurement and so in fact represents a general derivation of the uncertainty principle. This is because any two-level system can be modeled as a fictitious spin-1/2 system [48]; quantum transitions are always between two distinct levels. The evolution of the two states is well represented by a simple time-dependent vector in an appropriate Hilbert space, and is exactly analogous to a spin-1/2 precessing in a magnetic field in real space, with the energy level spacing playing the role of the magnetic field, and the weak oscillating perturbation in the role of the oscillating magnetic field. Such descriptions are widely used, for example, in quantum optics.

Instead of repeating the measurement N separate times, we can consider an ensemble of N uncorrelated systems measured simultaneously, and the same $\sqrt{N}$ reduction in frequency uncertainty results. This demonstrates the advantage of working with the highest possible density of systems (atoms, neutrons, etc.).

From (3.1), an EDM d in an electric field $\boldsymbol{E}$ gives a contribution to the precession frequency of $f_{\mathrm{e}} = 2dE/\hbar$, where we have assumed $F = 1/2$, leading to an EDM measurement uncertainty of

$$\delta d = \frac{\hbar}{2eET\sqrt{mN}}, \tag{3.5}$$

where, as before, the dimension of d is charge×length (e cm) and m represents the number of separate complete measurements of the N uncorrelated systems.

To define a figure of merit for a given EDM measurement technique, it is useful to rewrite the above equation in a slightly different form which takes into account the total measurement time t required to achieve a given sensitivity: $mN = It$ is the total number of systems measured, where I is an appropriate flux or counting rate. For example, in a neutron beam experiment, I is simply the number of neutrons per second which pass through the measurement apparatus. We are thus led to define a figure of merit M as

$$M = \alpha\beta ET\sqrt{I}, \tag{3.6}$$

where $\beta \leq 1$ takes into account inefficiencies such as imperfect polarization and dead time associated with electric field reversal, etc., and α relates the net system EDM to internal CP-odd physical processes. In certain special cases, for example, an electron EDM in a heavy paramagnetic atom, there can be an enhancement of a net atomic EDM by a factor of order 1000, as in the case of Tl. Also special is the case of polar molecule EDM experiments where a weak (polarizing) external electric field leads to an enormous internal (molecular) field. However, for an electron EDM in ^{199}Hg which is diamagnetic, $\alpha \approx 0.01$ due to shielding, as discussed in Chap. 6.

After a total time t of measuring ($m \approx t/T$), the limit on an EDM is

$$\delta d = \frac{\hbar}{2eM\sqrt{t}}. \tag{3.7}$$

For a storage experiment, where N particles are stored for a time T, the effective flux is $I = N/T$. In this case, it is interesting to note that

$$M = \alpha\beta ET\sqrt{N/T} = \alpha\beta E\sqrt{NT}, \tag{3.8}$$

and we see that the figure of merit increases only as the square root of the storage time.

3.1.2 Ramsey's Method of Separated Oscillatory Fields

Ramsey's method of separated oscillatory fields is a near realization of an ideal spin precession measurement, and is used in a number of EDM experiments. Ramsey was led to this idea through considerations of the Michelson stellar interferometer [49, 50].

Referring to the coordinate system shown in Fig. 3.1, in addition to the static field B_0, assume we can apply an oscillating magnetic field along the $\hat{y}$ axis for some fixed duration τ. Let us further assume that the oscillation frequency is tuned almost exactly to the Larmor frequency $\omega_0 = \gamma B_0$, where γ is the gyromagnetic ratio. The oscillating field is described by $2B_1 \cos\omega t\hat{y}$, where $|\omega - \omega_0| \ll \gamma B_1$.

The linear field can be decomposed into two counter-rotating field components;

$$B_1(\sin\omega t\ \hat{x} + \cos\omega t\ \hat{y}) + B_1(-\sin\omega t\ \hat{x} + \cos\omega t\ \hat{y}).$$

If we transform into a frame rotating at ω along $\hat{z}$, one of these fields will appear to be static while the other oscillates at 2ω. In most situations ($B_1 \ll B_0$, for example) we can ignore the counter-rotating field at 2ω; this is called the rotating wave approximation. The rotation direction is chosen to correspond to the spin precession direction; by Larmor's theorem, application of a magnetic field is equivalent to transforming into a frame which rotates at the Larmor frequency along the applied field, and in this case, in the rotating frame, the static field appears to be reduced; $B_0 \to B_0 - \omega/\gamma$.

In this frame, a spin originally along B_0 precesses around the y axis at a rate $\gamma B_1/2$, when the RF field frequency is perfectly adjusted to the Larmor frequency. Otherwise, the spin precesses around the vector sum of the residual z field, $\Delta B = B_0 - \omega/2\gamma$, and the y field, $B_1/2$.

Ramsey's method of separated oscillatory fields is a multi-step process which we will outline now. In this discussion, we assume $\Delta B \ll B_1$, and we can assume that when the RF field is applied, the spin precesses only about y (in the rotating frame).

1. The system is prepared with the spin pointing along z (B_0), and the RF field is off.
2. The RF field is turned on for a time τ such that $\gamma B_1\tau/2 = \pi/2$; this is called a $\pi/2$ pulse, and after it is applied, the spin lies along the x axis in the rotating frame. An important point is that it is perpendicular to the y axis (RF field) in the rotating frame.

3. After the RF field is turned off, the spin precesses very slowly in the rotating frame, at a rate $\Delta\omega = \gamma\Delta B$. The system is left to precess freely for a time T; this precession is in the $x - y$ plane, about the z axis. The spin angle, relative to x, is $\phi = \Delta\omega T$. The important point here is that if $\Delta\omega \neq 0$, after a time T, the spin is no longer perpendicular to y in the rotating frame.
4. A second $\pi/2$ pulse is applied. Only the spin component along x precesses, and after the pulse, this component lies along $-z$. The projection of the spin along x is given by $\cos\phi$.
5. An analyzer for the z-spin component is applied to the system; the difference in the probability of finding a spin pointing along $+z$ and along $-z$ is

$$P_+ - P_- = -\cos\phi = -\cos\Delta\omega T. \tag{3.9}$$

We thus have a way to determine the spin precession frequency relative to a stable oscillator. The oscillations in the spin direction (or spin flip probability) are sometimes called "fringes" in analogy to an interferometer.

An important feature of the separated field technique is that it measures the accumulated phase of the spin relative to the oscillating field; if $\Delta\omega$ varies over T, the final angle ϕ is given by the integral of $\Delta\omega$ over the time T.

When $\Delta B > B_1$, (3.9) no longer gives a good description of the separated oscillatory field process. This is because in this limit the $\pi/2$ pulses are no longer effective. However, our simplified picture gives a result in exact agreement with a more sophisticated treatment: in the limit $b = 2B_1 \gg \Delta\omega$, when $\pi/2$ pulses are used and $T \gg \tau$, equation (V.37) given in Ramsey's book [51] yields the same result.

In a beam experiment, the two $\pi/2$ pulses are given by two coils, through which oscillating currents are sent, and which are separated by a distance L along the beam propagation direction; the electric field plates are between the two coils and this zone is referred to as the interaction region. Beams usually have a wide velocity distribution, so that the interaction time $T \approx L/v$ is different for different atoms (or neutrons), and (3.9) must be averaged over this distribution. That is,

$$\langle P_+ - P_- \rangle = -\int \cos\Delta\omega T(v)\ N(v)\mathrm{d}v. \tag{3.10}$$

When $\Delta\omega$ is large, the average spin flip probability over the ensemble tends to zero and the "fringe visibility" is reduced; that is to say, the amplitude of the oscillations as given by (3.9) is reduced from unity. Also, the $\pi/2$ pulse is velocity dependent, with $\tau = \ell/v$, where ℓ is the coil dimension. Ramsey gives an extensive discussion of these points [51].

On the other hand, in storage experiments, the RF pulses (now separated in time) are generated by the same coil. Thus, the time T is the same for all members of the ensemble regardless of velocity; the same argument applies to τ, assuming that the RF field is homogeneous or that the storage volume is sufficiently averaged over the pulse duration by the moving spins.

3.1.3 Linewidth and Sensitivity with Separated Oscillatory Fields

When separated oscillatory fields are used to infer the spin precession frequency, determination of the effective line width is not so obvious. If the same apparatus is used both to produce the spin polarization before, and to analyze the $\pm\hat{z}$ components after, the RF pulses, the fraction of the total ensemble which is detected with spin in either direction, labeled as $\pm$ respectively, is

$$\frac{N_\pm(\Delta\omega)}{N_0} = (1 \mp \alpha \cos \Delta\omega T)/2, \tag{3.11}$$

$$\alpha = \frac{N_{\mathrm{max}} - N_{\mathrm{min}}}{N_{\mathrm{max}} + N_{\mathrm{min}}},$$

where $N_{\mathrm{max,min}}$ are the number of spins remaining in the original direction at a fringe maximum or minimum.

Setting $\Delta\omega = 0$ results in a quadratic sensitivity for ϕ; to achieve a linear sensitivity in storage experiments, the frequency is offset to the side of a fringe to the point where the sensitivity to frequency changes is highest. This point is approximately $\Delta\omega T = \pi/2$. In a beam experiment, it is easy to put a $\pi/2$ phase shift in the drive current to the second flip coil, or to orient the second coil field perpendicular to the first; in either case no frequency offset is needed. For both beam and storage experiments, the change in the spin-up minus spin-down fraction with frequency at maximum sensitivity is

$$\frac{\mathrm{d}(N_+ - N_-)}{\mathrm{d}\Delta\omega} = \frac{\alpha T}{2}(N_+ + N_-). \tag{3.12}$$

Since the standard deviation of $N_\pm$ is simply $\sqrt{N_\pm}$ and $N_+ + N_- = N_0$, the variance in $N_+ - N_-$ is $\sqrt{N_0}$, and therefore the standard deviation of $\Delta\omega$ is

$$\sigma(\Delta\omega) = \frac{2}{\alpha T \sqrt{N_0}}, \tag{3.13}$$

which is twice the Heisenberg limit obtained previously.

3.2 Ground State Optical Pumping and Detection of Atomic Polarization

Optical pumping of a ground state ensemble of atoms is widely used in atomic EDM experiments and it is therefore useful to describe the general principles here. An introduction to the subject is given in [52] and a complete review of the formal theory is given in [53]. As a concrete but simple example, we

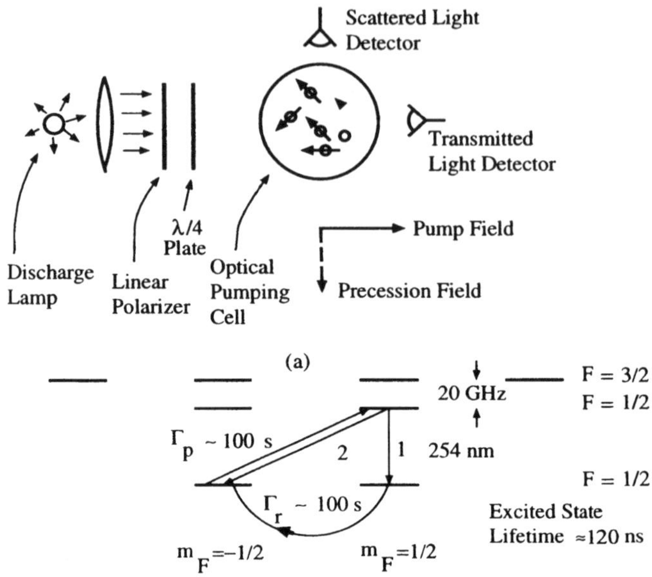

Fig. 3.2. Optical pumping of a spin-1/2 atom through the $F = 1/2$ excited state

can consider pumping an $F = 1/2$ ground state through an $F = 1/2$ excited state; this is in fact exactly the case used in the ^{199}Hg EDM experiment.

Consider the system shown in Fig. 3.2. Left-handed circularly polarized light propagates along an applied magnetic field and illuminates an ensemble of atoms. Since the light carries +1 unit of angular momentum, only atoms in the $m_F = -1/2$ level can absorb the incident light, and those atoms are transferred to the $m_F = +1/2$ excited state. The excited state lifetime is on the order of nanoseconds; the atom can either emit a photon or be quenched to the ground state in a buffer gas collision. In the either case, the atom has a nonzero probability of returning to the $+1/2$ ground state (the probability is 1/3 for both de-excitation processes). The important point is that some fraction end up in the $+1/2$ ground state and no longer absorb the resonance radiation, and eventually the entire sample becomes polarized along the light propagation (magnetic field) direction. During this process, the transparency of the vapor increases. The rate of pumping through the excited state (Γ_p) has to be fast relative to the rate of ground state polarization relaxation (Γ_r) due to, for example, interactions at the cell walls. It is useful to define a pumping rate as the number of photons scattered per second per atom in the ensemble; in equilibrium, the fraction of atoms polarized is then

$$P_\mathrm{a} = \frac{\Gamma_\mathrm{p}}{\Gamma_\mathrm{r} + \Gamma_\mathrm{p}}. \tag{3.14}$$

Furthermore, if the light is not perfectly circularly polarized, P_a will be reduced by a factor $P_L = (I_+ - I_-)/(I_+ + I_-)$ where $I_\pm$ refer to the left- and right-hand circular components of the light intensity.

Now if we suddenly switch the "pump" magnetic field off, and simultaneously apply a "precession" magnetic field perpendicular to the light propagation direction, the polarized atoms will begin to precess. If we continue to specify the quantization axis as the light propagation direction, we see the atomic population oscillate between the $m_F = \pm 1/2$ states at the Larmor frequency. The total relaxation rate of the ensemble polarization is $\Gamma = \Gamma_r + \Gamma_p$.

There are a number of ways to monitor this precession. One possibility is to place a magnetic pickup coil around the sample and measure the induced voltage. Also, if the optical pumping and precession is done using an atomic beam, then a Stern–Gerlach magnetic field can be used to monitor the final spin direction. However, the majority of modern EDM experiments use optical detection techniques. The atomic spin precession can be monitored either through the scattered resonance light intensity, or by simply measuring the transmitted light intensity [54]. The latter method has a much higher signal-to-noise ratio (in the case of one to two absorption lengths of vapor) because every scattered photon contributes to the signal. To achieve the same sensitivity by monitoring the scattered light, detectors would have to intercept a 4π solid angle around the cell. However, for cells where the absorption is low, or in low density atomic beams, collection of the scattered (fluorescent) light gives a higher signal to noise; this is simply because the fractional modulation for a low density system is very small, and the shot noise on the steady background light dominates the noise in a transmission measurement. The scattered light noise is determined by the counting statistics of useful photons.

The transmitted light intensity as a function of time, taking into account decay of the atomic polarization, is

$$I(t) = I_0 \exp[-\rho\sigma\ell(1 - P_L P_a e^{-\Gamma t} \cos\omega_0 t)] + I_b, \tag{3.15}$$

where ρ is the atomic density, σ is the resonance light absorption cross-section, ℓ is the cell length, I_0 is the intensity of light which is resonant with the atoms, and I_b is nonresonant background light (in this equation, σ represents an appropriately weighted average over the light spectral profile). The total process described above is known as a "pump–probe" technique; various EDM experiments described later use other optical pumping variants, but the basic mechanism is the same in all cases.

3.2.1 Atomic Spin Magnetometers

The above described properties of optical pumping lead us to formulate a general phenomenological description of the process applicable to any configuration of fields (including RF) and modulated pumping light.

First, to describe the interaction of the light with a polarized atomic ensemble, it is useful to introduce a vector to describe the light polarization; this vector, $\boldsymbol{P}_{\mathrm{L}} = P_{\mathrm{L}}\hat{k}$ lies along the light propagation direction $\hat{k}$ and $|P_{\mathrm{L}}| \leq 1$. As discussed before, if we describe the atomic polarization (dipole component) as a vector $\boldsymbol{P}$ with $|\boldsymbol{P}| \leq 1$, we find that the absorptivity of the ensemble is $\rho\sigma\ell(1 - \boldsymbol{P}_{\mathrm{L}} \cdot \boldsymbol{P})$, which reproduces the previously derived result (in the appropriate geometry).

The equations of motion (Bloch equations) for the atomic polarization can be now written down easily if we use the classical description of a magnetic moment in an arbitrary magnetic field

$$\frac{\mathrm{d}\boldsymbol{P}}{\mathrm{d}t} = -\gamma \boldsymbol{P} \times \boldsymbol{B} - \Gamma_r \boldsymbol{P} - \Gamma_{\mathrm{p}} \boldsymbol{P} + \boldsymbol{P}_{\mathrm{L}} \Gamma_{\mathrm{p}}, \tag{3.16}$$

where γ is the gyromagnetic ratio; two of the terms can be combined by defining $\Gamma = \Gamma_{\mathrm{p}} + \Gamma_{\mathrm{r}}$. The first term here is simply the Heisenberg equation of motion for a spin in a magnetic field; the other terms represent a phenomenological description of the pumping and relaxation processes. This equation is an almost exact description for a spin-1/2 system and gives a qualitative description for higher spin; for all practical purposes, this equation gives a fully adequate description of the evolution of the ensemble quantum state. Here we have neglected possible higher multipolarities in the description of the spin density matrix, but as we have discussed already, it is the dipole (vector) component that is of primary importance for EDM measurements.

Some experiments, to be fully described later, are based on changes in the equilibrium atomic polarization predicted by (3.16) on application of an electric field; an EDM can be included by adding an effective magnetic field which generates the same spin precession as an EDM in the applied electric field. We can thus describe all EDM measurements as equivalent magnetic field measurements.

In the case of weak magnetic fields (that is, $\omega_0 \ll \Gamma$), the steady-state solution of (3.16) is, assuming the pump light is along the z axis,

$$P_z = \frac{P_{\mathrm{L}}\Gamma_{\mathrm{p}}}{\Gamma} \gg P_y = \frac{\gamma P_z B_x}{\Gamma}, \quad P_x = -\frac{\gamma P_z B_y}{\Gamma}. \tag{3.17}$$

If we now include a weak probe beam along the y axis, a linear sensitivity to fields along the x axis is obtained; if the atom has an EDM, an electric field along x would rotate the ensemble polarization, primarily along z, and slightly along the y axis. The transmission through the vapor of a circularly polarized beam propagating along y would thus be changed. It should be noted that the probe beam will contribute to Γ; in principle, the probe beam should have zero average polarization. This technique was used in the Amherst Cs EDM experiment.

Now consider the case where a strong static magnetic field is applied. By application of a RF field $2B_1 \cos \omega t$, a steady spin oscillation can be established; in this case, a frequency shift is manifest as a phase change between the RF field and spin precession. Assume that B_0 lies along the $\hat{z}$ axis, the pump light lies in the x–z plane at an angle θ relative to $\hat{z}$. We are now assuming weak pumping light, $\Gamma_\mathrm{p} \approx \Gamma_\mathrm{r} \ll \omega_0$, and that the oscillating field is along $\hat{y}$. As before, we can transform into a frame rotating about the z axis, and in this frame, the effective magnetic field becomes $B_0 - \omega/\gamma$. Only the light polarization projected along B_0 will generate an atomic polarization. This is because if we transform into a rotating frame along B_0, the component of the light polarization vector parallel to B_0 is static, while the others now oscillate at the Larmor frequency. We can use the so-called secular approximation, as in celestial mechanics; the basic idea is that the effect of a weak driving term is significant only over a long time average. The effect of the oscillating polarization is reduced by a factor $\Gamma_\mathrm{p}/\omega_0 \ll 1$.

We are interested in the case of very weak effective fields in the rotating frame; defining $\Delta\omega = \gamma B_0 - \omega$, neglecting the counter-rotating component of the RF field, the steady-state solution to (3.16) is

$$\begin{aligned} P_z &= \frac{P_\mathrm{L}\Gamma_\mathrm{p}\cos\theta}{\Gamma}\frac{\Delta\omega^2+\Gamma^2}{\omega_1^2+\Gamma^2+\Delta\omega^2}, \\ P_y &= \frac{\omega_1\Delta\omega}{\Delta\omega^2+\Gamma^2}P_z, \\ P_x &= \frac{\omega_1\Gamma}{\Delta\omega^2+\Gamma^2}P_z. \end{aligned} \tag{3.18}$$

To calculate the absorption of the light, we can either transform the atomic polarization vector back to the laboratory frame, or transform the light polarization to the rotating frame; the latter is easier, and

$$\boldsymbol{P}_\mathrm{L} \to P_\mathrm{L}\cos\theta\hat{z} + P_\mathrm{L}\sin\theta(\hat{x}\cos\omega t - \hat{y}\sin\omega t).$$

Taking the dot product in the rotating frame, we find

$$\begin{aligned} \boldsymbol{P}\cdot\boldsymbol{P}_\mathrm{L} &= \frac{\omega_1 P_L\Gamma_\mathrm{p}\cos\theta\sin\theta}{\omega_1^2+\Gamma^2+\Delta\omega^2}\left(\cos\omega t - \frac{\Delta\omega}{\Gamma}\sin\omega t\right) \\ &\quad + \frac{P_\mathrm{L}\Gamma_\mathrm{p}}{\Gamma}\frac{\Delta\omega^2+\Gamma^2}{\omega_1^2+\Gamma^2+\Delta\omega^2}. \end{aligned} \tag{3.19}$$

The phase angle between the RF field and the oscillating component of the transmitted light intensity is

$$\tan\phi = \frac{\Delta\omega}{\Gamma} \approx \phi. \tag{3.20}$$

If an electric field is applied parallel to B_0, the effect of an EDM can be directly measured through the use of phase sensitive detection techniques. Maximum sensitivity occurs when $\Gamma_\mathrm{p} = 3\Gamma_\mathrm{r}$, $\omega_1 = \Gamma_\mathrm{r}$, and $\theta = 45°$. Note that this technique could also be implemented with two crossed beams, one to pump (along z), and the other to detect the precession in the x–y plane. This is more complex in that two light sources are required, in addition to extra optical paths through the apparatus. This technique was used in the first version of the ^{199}Hg EDM experiment.

As a final exercise, we will consider the case of modulated pump light along y with a static field along z. In this case, the light polarization in the laboratory frame is described by $\boldsymbol{P}_\mathrm{L} = \hat{y}P_\mathrm{L}\cos\omega t$, which, on transforming to a rotating frame at frequency ω around z, becomes (on average) $\boldsymbol{P}_\mathrm{L} = \hat{y}P_\mathrm{L}/2$. As before, the effective magnetic field in the rotating frame gives a precession frequency $\Delta\omega = \gamma B_0 - \omega$, and with (3.16) we find that the atomic polarization in the rotating frame is

$$P_y = \frac{P_\mathrm{L}\Gamma_p\Gamma}{(\Delta\omega^2 + \Gamma^2)}; \quad P_x = \frac{\Delta\omega}{\Gamma}P_y. \tag{3.21}$$

Transforming the atomic polarization back to the laboratory frame, $P_y \to P_y\cos\omega t$ and $P_x \to P_x\sin\omega t$. Taking the dot product between the light and atomic polarization vectors, we find

$$\boldsymbol{P}\cdot\boldsymbol{P}_\mathrm{L} = \frac{\Gamma_\mathrm{p}\Gamma}{4(\Delta\omega^2 + \Gamma^2)}\left(1 + \cos 2\omega t + \frac{\Delta\omega}{\Gamma}\sin 2\omega t\right), \tag{3.22}$$

and, similar to the previous case, there is a phase shift between the transmitted light modulation and the incident polarization modulation. Thus, this system, through an appropriate phase detection scheme, is sensitive to a shift in the Larmor frequency relative to the light modulation frequency.

The optimum pump rate for this system can be easily calculated. The shot noise on the transmitted photon flux is proportional to $\sqrt{\Gamma_\mathrm{p}}$, the sensitivity to a frequency shift is proportional to $1/\Gamma$, and the atomic polarization is Γ_p/Γ; we therefore must maximize

$$\frac{\Gamma_\mathrm{p}}{\Gamma_\mathrm{p} + \Gamma_\mathrm{r}}\frac{1}{\Gamma_\mathrm{p} + \Gamma_\mathrm{r}}\sqrt{\Gamma_\mathrm{p}}, \tag{3.23}$$

which occurs when $\Gamma_\mathrm{p} = 3\Gamma_\mathrm{r}$. We therefore see the figure of merit scale as the square root of the coherence time, as is usual for a storage experiment.

As a final note, in the case of one to two absorption lengths, the pump rate is given by the number of photons per second incident on the cell Φ divided by the number of atoms in the cell N, or

$$\Gamma_\mathrm{p} = \frac{\Phi}{N}. \tag{3.24}$$

The density is set such that $\rho\sigma\ell \approx 1$, and $N \approx \rho\ell hw$ where h and w are the transverse dimensions of the cell; furthermore, as we have shown, the pump rate should be approximately equal to the relaxation rate (or $1/T$, the coherence time). Thus, the photon flux is

$$\Phi \approx \frac{hw}{\sigma T}.$$

The figure of merit is given by the square root of the photon flux times the coherence time, which is

$$\sqrt{\frac{hwT}{\sigma}}, \tag{3.25}$$

and we see that a large cell is desirable (here we have assumed that the relaxation rate is independent of cell dimension; if the relaxation time increases with dimension, then there is an additional increase in the figure of merit with cell dimension), and also a weak atomic absorption which implies a greater number of atoms for one absorption length. The maximum electric field for a gas-filled cell is independent of h; this implies that as h increases, the applied voltage must also be increased to maintain the same sensitivity.

3.3 Electric Fields and Coherence Times in Various Systems

To set the scale of experimental possibilities and to provide a background for more detailed discussions later, we will review some basic features of the physical systems used or proposed for EDM measurements. The sensitivity of an experiment scales linearly with the electric field magnitude; attaining the maximum possible electric field in a given experiment is therefore of crucial importance. However, EDM experiments using systems (e.g., polar molecules) that can be fully polarized with a relatively small electric field are an exception to this.

3.3.1 Electric Fields in Vacuum

All beam experiments and some storage experiments must be performed under conditions of high vacuum (pressure less than 10^{-5} torr). The maximum electric field which can be stably maintained between two isolated plates (electrodes) with an open gap is of order 300 kV/cm; however, in most experiments, the field has to be periodically reversed so in fact the practical limit is of order 200 kV/cm [55]. Choice of electrode material is of crucial importance, as is the surface preparation. Generally speaking, the material must be highly polished, have a low vapor-pressure increase with temperature (e.g., a refractory material like molybdenum), and have low internally absorbed gas (e.g., copper). If the surface is rough, field emission takes place; if the electrode material has a high vapor pressure or is gassy, a spark between the electrodes can lead to a self-sustaining continuous discharge.

If dielectric insulators or spacers are placed between the electrodes in the high field region, the maximum stable field can be substantially reduced. Under such conditions, the maximum potential difference is about 300 kV for refractory materials, independent of electrode spacing [56]. Plastics or other materials which break down easily on heating or ion bombardment reduce the maximum field even further. It is also important that the contact area between the dielectric and the spacer be recessed into the electrode, as shown in Fig. 3.3 [57]. The field discontinuities at the sharp edges at the end of the dielectric will lead to electrical breakdown; this point, in regard to the

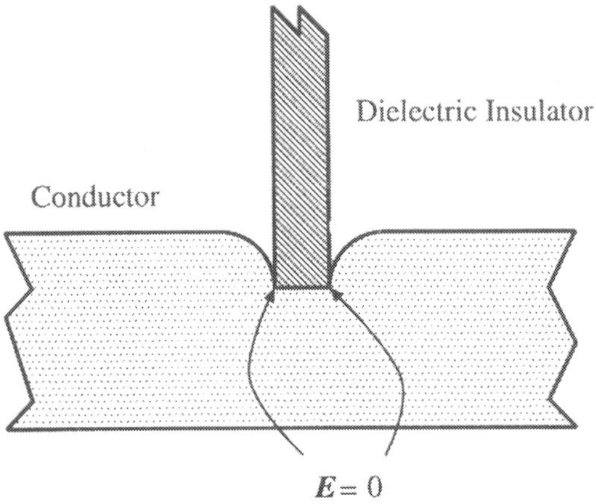

Fig. 3.3. Placing the edge of a dielectric insulator into a recess in a conducting electrode vastly improves the high voltage stability

neutron EDM experiment, has been addressed in [58], and the same ideas were important for the success of a neutron lifetime experiment based on the trapping of decay protons [59]. The idea of the recess is that the electric field is essentially zero in the corners of the groove, and the edge of the dielectric is effectively shielded. Alternatively, an appropriate bonding agent between the dielectric and conducting surface can be used, but the use of such materials usually requires a system compromise.

Systems where high voltage is applied in vacuum must be continually pumped because inevitable small sparks lead to outgassing and eventually the pressure will become too high; at a pressure greater than 10^{-2} torr, application of high voltage will result in a glow discharge in the gas. Interestingly enough, subjecting electrodes to a weak glow discharge can substantially improve the electrical stability properties under ultimate high-vacuum operation, and such treatment is in general referred to as "conditioning". The basic idea is that sharp points, dust, and organic dirt are destroyed in a glow discharge. An alternative conditioning technique is based on raising the electric field to a substantially higher value than will be used in the measurement; the sparking and discharging which occurs ablates away sharp points, dirt, etc. as in the case of a glow discharge. This procedure is often done with an increased background gas pressure. "Tuning" a capacitance in parallel with the electrode so that the sparks have sufficient energy to be effective yet not so powerful as to damage the dielectric and electrode surfaces is discussed in [58].

Sometimes a continuous background gas pressure of $10^{-5}-10^{-4}$ torr of an inert gas improves the high voltage stability; helium is particularly effective in this regard.

3.3.2 Electric Fields in Gases

Optical pumping of atomic vapors in sealed containers or "cells" is widely used for EDM searches; such systems represent a considerable simplification in, that, after the initial cell preparation, no vacuum system or pumping line is required, an important feature because the cells are generally buried deep in an apparatus. Since a high vacuum cannot be easily maintained in this case, the problem with runaway breakdown is avoided by purposely introducing a relatively high pressure insulating gas to the cell so that the inevitable slight pressure increase is irrelevant; N_2 is the most commonly used. In fact, optical pumping cells often include an inert gas (called a buffer gas), for example, to increase the diffusion time of atoms to the cell wall where spin disorientation occurs. Also, a buffer gas is important in systems where spin-exchange pumping is used.

At pressures greater than 1 torr (greater than the glow discharge regime), the dielectric strength of N_2 is about 50 V/torr. Helium and other noble gases are much less effective; apparently, to be an effective insulator, a gas must be slightly electronegative (an electron-eater) [60]. Typical gas pressures range from 300 to 2000 torr; electric fields are typically on the order of 10 kV/cm; at higher voltages, for cells of dimensions around 1 cm, leakage current effects and sparking along the cell walls become excessive.

3.3.3 Electric Fields in Liquids

No EDM experiments have been done with liquid solutions, although there was an early idea to perform an NMR experiment on an organic thallium compound. Electric fields in such a system would probably be limited to 10 kV/cm.

The idea of a neutron EDM experiment performed in a superfluid ^{4}He bath will be discussed later. Here it is expected that an electric field on the order of 500 kV/cm might be possible, due to the high dielectric strength of the liquid He, and to the reduction in surface leakage effects at low temperature.

3.3.4 Electric Fields in Solids

The discussion here is most relevant to experiments involving neutrons; this is because neutrons can easily pass through bulk matter, although the distance, hence precession time in the presence of the electric field, is limited due to the nuclear absorption rate. However, there has been a suggestion of an EDM experiment based on conventional NMR using atoms in a ferroelectric crystal. A very new possibility is the storage of alkali atoms in frozen helium; here fields of order $10^3 - 10^4$ kV/cm might be possible. This represents the highest dielectric strength typically available for any solid, an excellent example of which is polyethylene [57].

The use of ferroelectric crystals in a neutron EDM experiment has been suggested in [61]. These crystals exhibit spontaneous electric polarization analogous to the permanent magnetization of ferromagnets. A useful introduction to the subject is given in [62]. A polarized slab of a ferroelectric material could in principle produce an internal electric field of order 10^8 V/cm. Unfortunately, as mentioned above, solids do not have such a high dielectric strength and currents will invariably flow internally, leading to a build-up of free charges on the faces of the slab and a substantial cancellation of the internal field.

With neutrons, one has the possibility of sampling the microscopic fields in a noncentrosymmetric crystal. The average modulus of the electric field in a crystalline solid is of order 10^{10} V/cm, and a neutron experiment based on scattering in a CdS crystal had an effective field magnitude that was 10% of this value; this experiment will be described later. Fields of this magnitude or greater exist in diatomic molecules; molecular EDM experiments really comprise a separate subject.

The question of the stability of solid insulators under strong applied electric fields is important to essentially all EDM searches. This is because suitable materials must be used to support electric field plates and provide high-voltage vacuum feedthroughs, and, in the case of storage experiments, the storage cell normally comprises an insulating cylindrical shell separating two conducting plates. In this latter case, surface conductivity or leakage and corona effects limit the maximum fields to about 10–20 kV/cm; at higher fields, deviations from the current expected from the bulk conductivity increase rapidly and nonlinearly.

3.3.5 Coherence Times for Various Systems

The interaction time T in (3.9) is sometimes referred to as the coherence time; this is because T represents the time for which the spin system is in a coherent superposition. The velocities in beam experiments are on the order of 300 m/s; if the interaction region is about 1 m long, the effective coherence time is on the order of 3 ms. A new technique proposed to implement the old idea of an atomic fountain with ultracold Cs atoms might give up to 1 s coherence time.

The spin relaxation time (the maximum possible coherence time), for paramagnetic atoms in buffer gas storage cells is on the order of 10–100 ms, the limiting effect being spin disorientation in buffer gas collisions for short times or diffusion to the cell walls for the longer times. Diamagnetic atoms have coherence times in the range 100–4000 s and are limited by microscopic magnetic perturbations on the cell walls; such long times are obtained because the ${}^1S_0$ electron cloud effectively shields the atomic nucleus from perturbations during buffer gas or wall collisions, and the nuclear magnetic moment is 1000 times smaller than the electron moment so the coupling to external influences is substantially reduced.

For ultracold neutron storage experiments, the coherence time is limited by nuclear absorption and upscattering at the cell walls to about 100 s; the β decay lifetime of about 890 s is the ultimate coherence time. In a sufficiently homogeneous magnetic field, and when magnetic materials are avoided in construction of the storage cell, polarization relaxation of ultracold neutrons is not significant.

3.4 Magnetic Field Control and Generation

3.4.1 Field Stability and Homogeneity Requirements

As discussed already, an EDM of 10^{-26} $e\,$cm, under the influence of a 10 kV/cm electric field, corresponds to a spin precession frequency shift of 0.1 μHz, and implies an effective magnetic field of about 0.1 nG for a diamagnetic atom, or 0.1 pG for a paramagnetic atom. Since an EDM would reveal itself through a frequency change on reversal of an applied electric field and there is some time delay between subsequent measurements, the net magnetic field must be stable over this time interval at the level of nG, or one part per billion of the Earth's field. Unfortunately, ambient magnetic fields are well known to fluctuate at the 10 mG level due to solar storms, human activity, etc.

Also important are field gradients which can cause dephasing between various regions of the cell in a storage experiment. The maximum tolerable size of a field gradient can be easily estimated by the following simple argument. The cell can be divided into two half-volumes with a relative magnetic field difference of δB, and a polarized atom (or neutron) moves randomly between the two halves by diffusion with an average time τ. The spin angle then undergoes a random walk as the atom moves between the two halves; it does this T/τ times in time T, each time acquiring a phase change $\pm\gamma B\tau$, which leads to an rms angular spread of

$$\phi - \gamma\delta B\tau\sqrt{\frac{T}{\tau}} = \gamma\delta B\sqrt{T\tau}. \tag{3.26}$$

If we assume that one relaxation lifetime corresponds to $\phi = \pi/2$, then

$$\Gamma = \frac{1}{T} = \frac{4}{\pi^2}(\gamma\delta B)^2\tau, \tag{3.27}$$

where γ has units rad/(s·G). This agrees well with a more sophisticated calculation [53]. For the ultracold neutron EDM experiment

$$\tau = L/v = 20\text{ cm}/500\text{ cm/s} = 0.04\text{ s} \tag{3.28}$$

and to achieve $\Gamma = 1/500$ s required $\delta B < 20$ μG, or a gradient of about 1 μG/cm. In the case of the ^{129}Xe EDM experiment where the cell is of order 1 cm, $\tau \approx 1$ s due to the diffusion time in the buffer gas, so $\delta B < 10$ μG, corresponding to a gradient of < 10 μG/cm.

3.4.2 Magnetic Shields

At one stroke, magnetic shields eliminate problems due to the Earth's field, 60 (or 50) Hz power line fields, magnetic storms, human activity, and field gradients. The development of magnetic shielding in regard to general EDM experimental technique is a great technical triumph. At present, external field fluctuation effects are not a limiting factor; the limiting systematic effects are internal to the apparatus and development of possible improvements, such as superconducting shields, really is not warranted at this time. The most amazing feature of magnetic shielding, in its present form, is that it is nearly completely passive yet can give up to a factor of 10^6 reduction in external magnetic perturbations; additional active compensation usually gives only incremental improvement but is also generally quite simple. Most experiments also employ magnetometers placed near the region where the actual EDM measurement takes place and can be thought of as an additional magnetic shield, in this case quite active, but primarily aimed at the measurement of internally generated fields. An ultimate goal for all EDM experiments is the development of an appropriate magnetometer that occupies the same volume, over the same precession time interval, as the species on which an EDM is sought. Such a magnetometer is now generally referred to as a comagnetometer [63]. However, as we will see again and again when specific experiments are described, there is often a price to pay on the inclusion of a second polarized species. The problem is that a spin–spin interaction can occur, and this results in an effective magnetic field that does not scale as the ratio of magnetic moments, expected for a true external magnetic field. Although the effective field does not necessarily lead to systematic effects, its presence can be a source of instability and extra noise.

The idea of using ferromagnetic materials for magnetic shielding is literally hundreds of years old. Nowadays there are special alloys for this purpose; one of the first commercially available was "Mu-Metal", and although it has a very high permeability, is very susceptible to work hardening by mechanical stresses and therefore a reduction in permeability. For a material to be useful for shielding, it must be magnetically "soft" (easily magnetized and demagnetized) and have a very high permeability. At present, the favored materials are "Moly Permalloy" and various Conetic alloys, with μ of order 20 000.

A useful review of the field with shielding factor approximations for various configurations is given in [64] and [65]. Figure 3.4 shows the effect on an initially homogeneous magnetic field for two different orientations of a magnetic shield. The first case is an infinitely long cylindrical shell with its axis transverse to the field. It is easy to show that the transverse shielding factor is (see [66], Problem 5.8)

$$S_T = \frac{B_i}{B_0} = \frac{\mu t}{2R}, \tag{3.29}$$

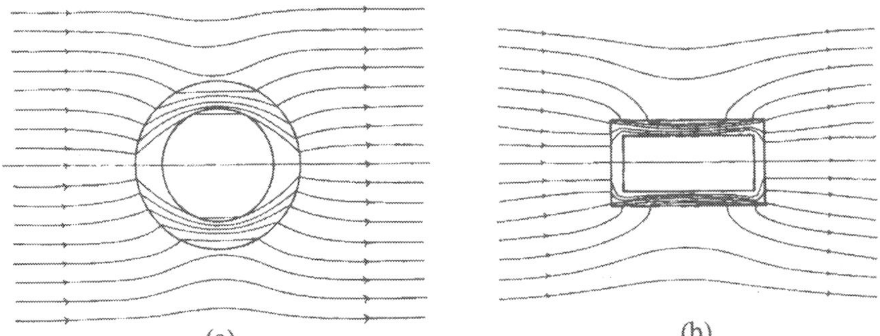

Fig. 3.4a,b. High-permeability shields in an initially homogeneous field. (a) Infinitely long cylinder with axis perpendicular to the applied field. (b) Short closed cylinder with axis parallel to the applied field

where B_0 is the homogeneous magnetic field before introducing the shield, B_i is the field inside the shield, μ is the permeability, t is the shell thickness, and R is the radius.

The axial shielding factor is a more complicated problem; for an infinitely long cylinder with its axis parallel to the homogeneous field, there is no shielding factor. Cylinders of more reasonable length have an axial shielding factor approximately

$$S_{\mathrm{A}} \approx \frac{2\mu t R^{1/2}}{L^{3/2}} \tag{3.30}$$

for cylinder length L in the range $4 < L/R < 80$. We thus see, in the axial case, the possibility of substantial reduction in shielding as compared to the transverse case.

A single layer shield is not adequate for most EDM work, hence the use of multilayer shielding, as shown in Fig. 3.5. Usually, the ends of such shields are in the form of removable lids; the overlap between the rim of the lid and cylindrical tube should be at least 10 cm. Also, the endcaps should fit well with very little gap (but not so tightly as to cause stresses). If these conditions are satisfied, the shield can be thought of as continuous for the purposes of calculations.

Holes for optical paths and electrical feedthrough degrade the shielding factor only slightly. For a hole of radius r in a shield with radius $R \gg r$, the external field perpendicular to the surface (the only component for a large shield of high permeability material; this is because the external flux lines enter perpendicular to the neighboring surface), the field inside the shield falls off as

$$B(h) = B_{\mathrm{ext}} \mathrm{e}^{-kh/r}, \tag{3.31}$$

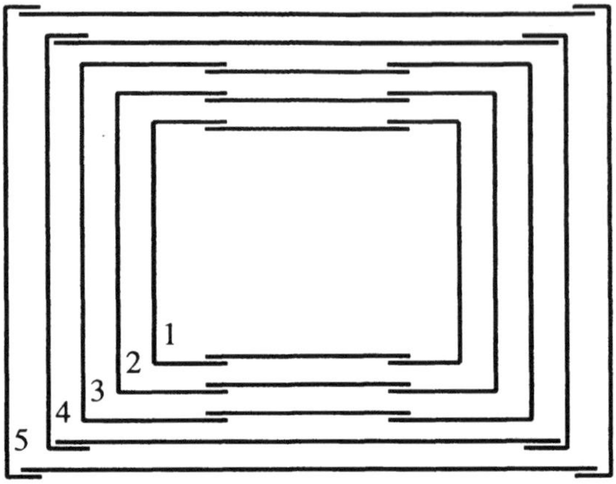

Shield	D(m)	L(m)	t (mm)
1	1.00	1.404	2.00
2	1.636	1.636	2.00
3	1.934	1.934	2.00
4	2.300	2.300	1.57
5	2.740	2.740	1.57

Fig. 3.5. A schematic cross-section view of the axially symmetric magnetic shields used in an ultracold neutron EDM experiment

where h is the distance from the hole and $k \approx 1.5$ [65]. For a series of concentric holes in a multilayer shield, the propagation of an external field into the shield interior can be better modeled as the propagation of a field into a high permeability pipe, with a substantial improvement over the above result. But a worst-case estimate can be obtained by simply taking a product of the effective shielding with h taken as the difference in radii between the layers.

The shielding factor for a multilayer shield has been estimated by a number of authors; an overview along with some particularly compelling results is presented in [64]. The net shielding factor is not simply a product of individual factors because the presence of the internal shield affects the internal boundary conditions. As shown in [64], the transverse shielding factor in the limit of large S_T^i is

$$S_{\mathrm{T}} = S_{\mathrm{T}}^n \prod_{i=1}^{n-1} S_{\mathrm{T}}^i \left[1 - \left(\frac{R_i}{R_{i+1}}\right)^2\right], \tag{3.32}$$

where the subscript i refers to the shielding factor, radius, etc. of the individual layers. Physically, the factor $1 - (R_i/R_{i+1})^2$ reflects the volume loss (compression of internal flux) into the region between the shield layers. The axial shielding is

$$S_{\mathrm{A}} = S_{\mathrm{A}}^n \prod_{i=1}^{n-1} S_{\mathrm{A}}^i \left(1 - \frac{L_i}{L_{i+1}}\right) \tag{3.33}$$

and is analogous to the previous equation; only now the reduction in volume is given by the relative lengths of the cylinders.

Shields as described here are usually made of rolled and welded sheet metal. After fabrication, they must be carefully annealed to ensure that there are no work-hardened, hence magnetically hard, regions. This is necessary to achieve the very high inherent material permeability. Under normal use, for example, when the shields are disassembled and reassembled, or if the internal field currents are significantly changed, the shield assembly must be "demagnetized" or "degaussed". This is accomplished by applying an alternating magnetization to the shields, with this magnetization established by several current loops through the shield, entering through holes in the endcaps and along the axis, looping around the outside. To magnetically saturate the shields, about 50 ampere-turns is required. The demagnetization is accomplished by alternating the current through the current loops, hence the magnetization, and gradually reducing this from the initial conditions of saturation to zero. The idea is that the magnetization, after saturation, is symmetric on current reversal. If the rate of current decrease is much less than the frequency of reversal, on reducing the current amplitude to zero, no net magnetization remains.

When internal fields are applied, the shields gradually drift as the magnetization reaches a new equilibrium value. Tapping on the shields or support assembly helps establish this equilibrium; a mechanical vibration helps the domains flip into their equilibrium state. It was recently discovered that performing the demagnetization with the internal fields applied achieves the equilibrium magnetization immediately and a major source of frequency drift is eliminated [67]. Tapping in this case is also important.

After demagnetization, the internal field gradients are quite small. The dominant gradient contributions are imperfections in the material (hard spots) and welds. These imperfections can be modeled as small current loops; we thus expect that residual gradient effects will scale as $1/R$, the inverse of the internal shield radius; this is because the total number of current loops increases as the area, but the gradient effects fall off as $1/R^3$. In fact, this is experimentally observed, and typically the internal gradients are 100 μG$/R$ with R measured in cm. We thus see that the required field homogeneity with respect to background gradients is easily achieved.

There are a few active techniques that give reasonable improvement to the effectiveness of passive shields. A flux-gate can be used in a feedback system to keep the external field (along the same axis as an internally applied field) at the shield surface fixed. The dynamic shielding factor can also be improved by "shaking". An oscillating magnetic field is applied with loops similar to those used for demagnetization; however, in this case, the current oscillation frequency is rather high (a few hundred Hz; this is set by the skin depth of the material). The oscillating field moves the internal magnetization along the hysteresis curve, and the domains are continually shaken into their equilibrium state. There is an optimum value for the oscillating field amplitude, and the shaking really cannot be applied to the innermost shield layer because of the perturbing influence of the leakage field.

For large diameter shields, it is likely that the outermost layer is near saturation, in which case the dynamic shielding is substantially reduced. However, for all ferromagnetic materials, there is a maximum dynamic permeability for some applied field, roughly one-half of the saturation field. This is because the slope of the hysteresis curve does not have its maximum at $B = 0$. Intentionally applying a field can only be used with the outermost layer and in some sense is equivalent to shaking.

3.4.3 Field Generation

The benefits in the use of magnetic shielding in optical pumping experiments were first demonstrated and discussed in [68]. It was discovered that a solenoidal winding on the internal surface of the shield cavity has a much better field homogeneity than the same winding in free space. It turns out that this represents a general principle for the generation of very homogeneous fields within a high-permeability shield [69].

The questions are: What distribution of currents on the internal surface of an infinite permeability shield will result in a perfectly homogeneous magnetic field, and is such a current distribution possible? The answer to the question is alluded to in Problem 17, Chap. IV of [70].

Any linear current of arbitrary size and configuration can be decomposed into a system of infinitesimal elementary currents, as shown in Fig. 42 of [70]. Now the magnetic moment of each loop is $\mathrm{d}\boldsymbol{m} = I\boldsymbol{n}\mathrm{d}a$ where $\boldsymbol{n}$ is a surface normal, and in general $I = I(\boldsymbol{x})$ where $\boldsymbol{x}$ specifies a location on the surface. Since we are considering the case of infinite permeability, the normal component of $\boldsymbol{B}$ must be continuous across the surface, while $\boldsymbol{H}$ is zero within the material and $\boldsymbol{B} = \boldsymbol{M}$, with $\boldsymbol{M} = \mathrm{d}\boldsymbol{m}$ for the elementary current loop. This problem can be solved by the use of potentials by noting that the effective surface charge is given by $\sigma = \boldsymbol{n} \cdot \boldsymbol{M} = I\mathrm{d}a$. We want the internal magnetic field to be homogeneous, that is, of constant magnitude along a fixed direction, described by the vector $\boldsymbol{B}_0$; since $\boldsymbol{M} = \boldsymbol{B}$ within the material, we see immediately that the surface charge to generate $\boldsymbol{B}_0$ is given by $\sigma = \boldsymbol{n} \cdot \boldsymbol{B}_0 = I(\boldsymbol{x})\mathrm{d}a$, or

$$I(\boldsymbol{x}) = \boldsymbol{n}(\boldsymbol{x}) \cdot \boldsymbol{B}_0 \,. \tag{3.34}$$

For a solenoid wound on the inner surface of a shield, the magnetic field will be perfectly homogeneous; the windings generate a constant $\boldsymbol{B}_0$ along the cylinder axis and the effective charge on the endcaps perfectly terminates the field lines. This can be seen as follows. Again, the windings can be decomposed into infinitesimal current loops; the decomposition is divided between the two shield ends at the axial midpoint, and on the endcaps, there are constant surface charges, equal but of opposite sign. The surface charge as a function of position along the shield axis varies linearly from one end-cap to the other. The potential for this charge distribution is a linear function of the axial coordinate, with no radial or azimuthal dependence. The field is thus perfectly constant and only along the axial direction. In this case, the same result can be obtained by considering image currents, which give an infinite solenoid.

A perfectly homogeneous transverse field can be generated by a "cosine" distribution winding, as shown in Fig. 3.6; one can see in this case that the distribution of infinitesimal current loops mimics the required $I(x)$. Of course, since the wires are of finite size and spacing, this only represents an approximation, but the deviations from a perfect field become vanishingly small for any reasonable number of turns.

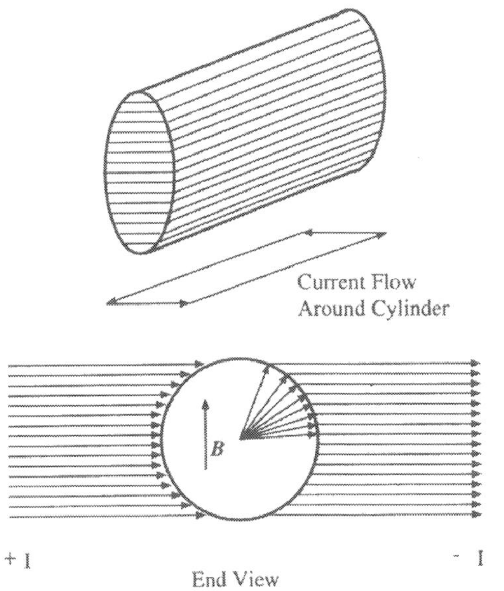

Fig. 3.6. A cosine distribution can be approximated by a network of wire loops. Each loop gives a constant effective surface charge, hence the surface charge maxima and magnetic poles, are in the low wire density regions

Of course, to achieve high stability in the applied field, very high quality current supplies must be used. Given that the field stability should be on the order of nG to pG, and typical fields of around 1 mG/mA, we see immediately that the current supply must have an absolute stability, at the level of 1 mA, of 1 pA; for the neutron EDM experiment where the Larmor frequency is 30 Hz, this corresponds to a fractional stability of over 1 part per billion. Nowadays, with high performance operational amplifiers, achieving this level of stability is straightforward.

Since the axial shielding is considerably worse that the transverse shielding, for experiments where an applied magnetic field is used, there is considerable advantage to having this field transverse to the shield axis; the axial leakage field adds in quadrature with the applied field. Also, holes in the shield have the least perturbing effect when placed at the poles of the internal field; the fewest number of windings need to be disturbed in this case, and although this is where the effective surface charge is maximum, the return flux in the shield material is minimum. These two points were crucial in regard to the ultracold neutron EDM experiment [71].

A useful technique which allows generation of very high axial fields is the nested solenoid geometry [72]. This technique can be applied coils in various geometries and is widely used in electrical engineering to reduce the leakage flux of inductors (see [73], Sect. 2, §22). The idea is that the leakage is represented by the total flux a coil generates; in a magnetic shield, this flux must return through the ferromagnetic material and if the field is large, can lead to saturation, or, if the field must be switched on and off or reversed, a remnant magnetization, which decays slowly, is generated. For the case of nested solenoids, two coaxial windings are made, and currents are sent through the coils in opposite direction such that the magnetic flux is $\Phi = B_1 A_1 = -B_2 A_2$ where A represents the cross-sectional area of a solenoid, and B is the field. Alternatively, the number of turns per inch for each coil can be adjusted so that the currents in each are the same while still canceling the net flux. The net field inside the inner solenoid is reduced to $B_1 \to B_1 - B_2 = B_1(1 - A_1/A_2)$ so there is only a modest reduction in field strength. Without such cancellation, field fluctuations due to changing properties of the shield material (i.e., temperature variation of the permeability or saturation characteristics) can be excessive in the case of a strong internal field. If a weak field is changed, excessive drift due to the relaxation of the remnant magnetization can be a problem, and nesting can be helpful in this case also.

Given the perfect magnetic shielding of superconducting material compared to the finite shielding of ferromagnetic materials, there is some motivation to develop superconducting shields. However, the boundary conditions for a superconductor (a perfect diamagnetic material) are different from those of a high permeability material; the homogeneity of fields generated by current loops is worsened in the case of a superconducting shield and this is a contraindication for its use as the innermost shield layer. So far, all attempts to make large-scale superconducting shields have led to disappointing results. Some flux loops invariably remain trapped inside the shield, and the trapped flux varies (in sudden jumps) randomly. The field noise tends to be considerably greater than that from a well-demagnetized ferromagnetic shield.

3.5 Systematic Effects

As we have already mentioned, the art of seeking an EDM relies on the separation of a possible true EDM from spurious magnetic and other effects associated with the application of the maximum possible electric fields. Each experimental technique has its own set of problems; however, some problems are common to all methods.

For example, a changing magnetic field associated with the apparatus used to reverse the applied electric field polarity has led to systematic effects in a number of experiments. Such effects can be identified, for example, by reversing the leads to the power supply, but unless one is careful, an effect such as this can be very troublesome. Other systematic effects are not so easily contended with. Here we will give an overview of some of the more important ones.

3.5.1 Leakage Current Effects

Whenever high voltages are applied to a system, small leakage currents invariably flow through insulators, and these currents generate magnetic fields which are correlated with the electric field direction and are indistinguishable from an EDM. The leakage current magnetic field is a function of the electric field and adds a term to the Hamiltonian (3.1)

$$H = -\left[\mu(\boldsymbol{B} + \hat{z}(\beta\hat{z}\cdot\boldsymbol{E}) + d\boldsymbol{E}\right]\cdot\boldsymbol{F}/|F|\,, \tag{3.35}$$

where β represents the average projection of the magnetic field generated by the current density

$$\boldsymbol{j} = \sigma\boldsymbol{E} \tag{3.36}$$

along the static magnetic field direction ($\hat{z}$), with σ representing the electrical conductivity. A nonzero β implies some helicity of $\boldsymbol{j}$ along $\hat{z}$.

The apparent T-odd character of this new term is the result of the irreversible "macroscopic" process(es) which lead to (3.36). Also, under parity reversal, the helicity of the leakage path, hence the effective magnetic field, changes sign; under parity reversal, $\boldsymbol{E}$ also changes sign, so we see that the leakage current effect is even under parity reversal, unlike a true EDM. This and related points were discussed in Sect. 1.4.

For beam experiments where the insulators and conductors leading to the high voltage plates can be relatively well spatially separated from the sensitive measurement area, leakage current fields are not so troublesome as in the case of storage experiments where the cell walls generally serve a second purpose as the high voltage electrode spacer.

A worst-case scenario is when all the leakage current flows in a closed loop around the cell; given that such a current flow is highly unlikely, to estimate a possible systematic effect, one-quarter of the field at the center of a loop is sometimes taken. Note that if such a helical current existed due to some imperfection in the cell walls, reversing the cell orientation does not distinguish this effect because the leakage current helicity and hence magnetic field direction is a fixed property of the cell.

For essentially all cell-type EDM experiments, the cell is in the form of a pillbox, with the endcaps being conductive and forming the electrodes. The cylindrical shell wall is made of a high dielectric strength material; pyrex, fused silica, or beryllium oxide, to name a few. These materials have electrical resistivities in the range of 10^{14}–10^{16} Ω·cm. Typically, the diameter to height ratio of a cell is 2:1, and the wall thickness is 5% of the diameter. Also, most cell experiments use electric fields of about 10 kV/cm; we can therefore derive a rule-of-thumb systematic leakage-current effect which depends only on the radius of the cell R, assuming a typical resistivity of 10^{15} Ω·cm. The total cross-sectional area of the base of the cylindrical cell wall of thickness t is $a = 2\pi Rt = 0.05\times 2\pi R^2$. The total current which flows is $I = a\sigma E \approx 3R^2$ pA. The systematic magnetic field is then

$$B_{\rm sys} = \frac{1}{8}\frac{\mu_0 I}{2R} \approx 0.3R \text{ pG/cm}. \tag{3.37}$$

For a diamagnetic atom or neutron, the gyromagnetic ratio is about 1 kHz/G, so this corresponds to an systematic EDM of about $3\times 10^{-28} R\, e\,\text{cm}$, or for a paramagnetic atom where the gyromagnetic ratio is 1 MHz/G, a systematic EDM of order $3\times 10^{-25} R\, e\,\text{cm}$. These estimates, as will be discussed later, are quite close to the actual limits for various experiments. However, surface leakage currents are often as large as or bigger than the bulk currents. Even so, the rule-of-thumb estimate given above is rarely off by more than a factor of 10.

It is also important that the leads which supply the high voltage are coaxial with the leakage current return leads, otherwise the leakage currents, charging displacement currents when the electric field magnitude/direction is changed, or impulse currents associated with sparks can cause a systematic magnetization of the shields.

3.5.2 Problems Related to Polarizability and Electric Quadrupole Moments

For systems with total $J = 0$ or 1/2, the internal (m_J) states remain degenerate under application of an electric field. Thus, even though the neutron has a finite polarizability, application of a constant electric field will not directly affect its spin precession frequency. Also, atoms with ${}^2P_{1/2}$, ${}^2S_{1/2}$, or ${}^1S_0$ electronic configurations are not split by an electric field. We thus immediately see the advantage of working with spin-1/2 systems; such insensitivity to a vector polarizability is described in the context of Kramers' theorem which is really a statement about time reversal symmetry ([15], §60).

Unfortunately, atomic systems are not totally immune to electric field effects. In optical pumping experiments, the optically excited state can have a different scalar polarizability and hence, on application of an electric field, the optical transition frequency can change, even for transitions between $F = 1/2$ and $F = 1/2$ states. The order of magnitude is typically 100 MHz/(100 kV/cm)^2; at 10 kV applied field, this is about 0.1% of the transition linewidth and can in principle lead to an equivalent change in the atomic absorptivity, hence a systematic effect. The danger is that the electric field distribution can be polarity dependent; it is the average of E^2 over the interrogation region which is important. This point will be discussed later in relation to specific experiments.

If the nucleus of an atom has a quadrupole moment (which requires a nuclear spin $I > 1/2$), there can be an interaction with gradients in the applied electric field; a typical nuclear quadrupole moment is on the order of $10^{-24}\ e\,\mathrm{cm}^2$ or less; a gradient of 10 kV/cm^2 could cause a frequency shift of about 2×10^{-6} Hz. Note that the assumed gradient is a worst-case estimate, and systematics are generated through field-reversal asymmetry. Also, there is a question of how the quadrupole shift affects the magnetization precession frequency; our vector model of a precessing spin is not applicable in this case, and one must consider the higher moments of the density matrix. Most experiments, however, employ atoms with $I = 1/2$ nuclei; the exception is Cs, which has a particularly small nuclear electric quadrupole moment of about $10^{-26}\ e\,\mathrm{cm}^2$.

For $J = 0, 1/2$ electronic states, if $F > 1/2$ ($\boldsymbol{F} = \boldsymbol{I} + \boldsymbol{J}$ is the total atomic angular momentum), there can be an atomic quadrupole moment induced by the hyperfine interaction; this quadrupole is likely to be larger than the nuclear one. For Tl, it can be estimated:

$$\frac{E_{\mathrm{hfs}}}{E(p_{1/2}) - E(p_{3/2})} e a_0^2 \approx \frac{3\mathrm{GHz}}{3 \times 10^{14}\mathrm{Hz}} \times 10^{-16}\ e\,\mathrm{cm}^2 \approx 10^{-21}\ e\,\mathrm{cm}^2 . \tag{3.38}$$

A related problem is an effective modification of the apparent ground state magnetic moment under application of an electric field. If the ground and an excited state (of opposite parity) have different g factors, the energy splitting between the ground and excited states is different for $+m$ and $-m$ Zeeman sublevels. This leads to a slightly different scalar polarizability for the $+m$ and $-m$ ground states, resulting in an energy difference between the ground state sublevels which scales as the magnitude of the applied magnetic field B times the scalar polarizability shift αE^2, divided by the energy splitting between the opposite parity levels:

$$\frac{g_p - g_s}{E_p - E_s} \alpha E^2 \approx \frac{1\ \mathrm{MHz}}{(100\ \mathrm{kV/cm})^2} \tag{3.39}$$

for Tl or Cs. For a $^1S_0$ ground state, there is a further reduction factor given by the excited state hyperfine shift divided by $E_p - E_s$. This is because in the absence of an excited state hyperfine splitting, the effect vanishes in the case of a $^1S_0$ ground state electronic configuration.

3.5.3 The $\boldsymbol{v} \times \boldsymbol{E}$ Problem

Another spurious effect is the so-called motional magnetic field, first addressed in relation to a Cs EDM experiment. Its effects are most severe for atomic beam experiments [74]. When one moves relative to the sources of a static electric field $\boldsymbol{E}$, according to special relativity, a magnetic field $\boldsymbol{B}_{\mathrm{m}}$ is generated in the comoving frame which to first order in v/c is

$$\boldsymbol{B}_{\rm m} = \frac{\boldsymbol{v}}{c} \times \boldsymbol{E}. \tag{3.40}$$

For a typical thermal atomic velocity of $v = 300$ m/s in an electric field of 10 kV/cm, $B_{\rm m} = 30\ \mu$G. Consider an experiment where there is a large applied magnetic field $\boldsymbol{B}_0$ and an EDM is sought by measuring the shift in Larmor precession frequency on reversal of a electric field $\boldsymbol{E}$, as implied by (3.1). If

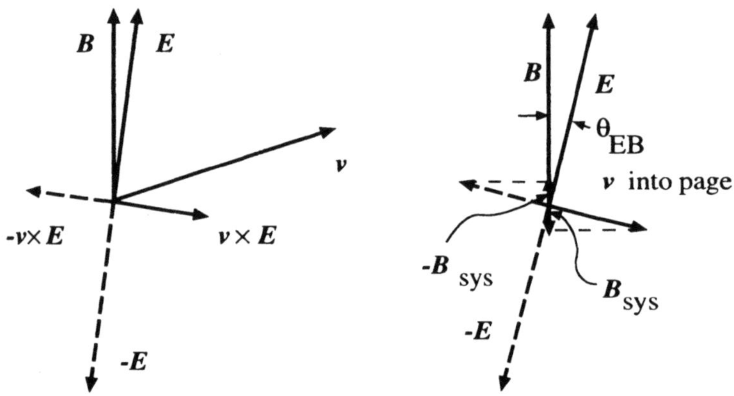

Fig. 3.7. Geometrical picture of the $\boldsymbol{v} \times \boldsymbol{E}$ effective magnetic field

$\boldsymbol{E}$ and $\boldsymbol{B}_0$ are nearly parallel as shown in Fig. 3.7, and $B_{\rm m} \ll B$, the effective magnetic field strength is given by the magnitude of $\boldsymbol{B} = \boldsymbol{B}_0 + \boldsymbol{B}_m$

$$B = B_0 + \theta_{\rm EB} B_{\rm m} + \frac{1}{2}\frac{B_{\rm m}^2}{B_0}, \tag{3.41}$$

giving a change in Larmor frequency of

$$\Delta\omega = \frac{\gamma\theta_{\rm EB} v}{c} E + \frac{\gamma v^2}{2c^2}\frac{E^2}{B_0},$$

where $\theta_{\rm EB}$ is the angle between $\boldsymbol{E}$ and $\boldsymbol{B}_0$ in the plane perpendicular to $\boldsymbol{v}$. If B is substituted into (3.1), it can be readily seen that if $\theta_{\rm EB} \neq 0$, a spurious shift correlated with $\boldsymbol{E} \cdot \boldsymbol{B}$ is generated, which has the same signature as an EDM. However, this is not a true T violating effect, for under T, $\boldsymbol{v}$ reverses sign, and therefore so does $\boldsymbol{B}_{\rm m}$. The important point is that reversing $\boldsymbol{E}$ relative to $\boldsymbol{B}$ does not create the time-reversed Hamiltonian; $\boldsymbol{v}$ must also be reversed. Even in the case where $\theta_{\rm EB} = 0$, there is a relative shift quadratic in $B_{\rm m}$, which may require that the magnitude of E does not change significantly on reversal. The $v \times E$ systematic was the ultimate limiting factor for neutron beam experiments. For atomic beam experiments, there are some tricks to get around the problem, as will be discussed later.

Although the motional field is most significant in the case of beam experiments, examples of which are the early neutron EDM experiments and the more recent thallium EDM experiment [75], there can be some subtle effects in other cases. EDM experiments using optically-pumped atoms in a cell have on average $\boldsymbol{v} = 0$ simply because the atoms are free to rattle about the cell, so one might expect that there is no net motional effect. However, as we will show, the fluctuating field associated with the random velocity can in fact lead to sizable systematic effects; the quadratic term in the effective magnetic field persists even if the average velocity is zero, and one may wonder why it is possible to measure EDMs to the achieved levels of sensitivity; if we consider a case where $B_0 = 10$ mG, and $v = 120$ m/s, the quadratic term amounts to about 50 nG, corresponding to a shift of 35 μHz in the case of optically pumped ^{199}Hg where the current experimental accuracy is at the level of 10^{-9} Hz (1 nHz), which implies a magnetic field of about 1 pG. This would require an electric field magnitude reversal symmetry of 1 part in 10^5.

An important point has been neglected in this estimate. In fact the motional magnetic field is randomly fluctuating, and it simply is not correct to take the average square of this field. The motional field has a definite magnitude only for a time interval τ_c, the time between substantial velocity changes due to, for example, collisions with buffer gas molecules or cell walls. The parameter τ_c depends on the system geometry, nature of the collisions, and velocity of the particles.

For a spin-1/2 system, the net effect of the randomly fluctuating field can be readily quantitatively calculated in the context of the density matrix. The Hamiltonian can be separated into static and time-dependent components

$$H = H_0 + H(t) = -2\pi\gamma\sigma_z B_0/2 - 2\pi\gamma f(t) B_m \sigma_x/2, \tag{3.42}$$

where γ is the gyromagnetic ratio (Hz/G), $\sigma_{x,z}$ Pauli matrices, and $f(t)$ represents the fluctuating character of B_m. Here we only consider the possibility of an x component of B_m, but this doesn't change the result significantly. Eventually, both time and ensemble averages of the effect of this Hamiltonian must be determined.

By transforming into a rotating frame, the static component of the Hamiltonian can be eliminated

$$H'(t) = e^{i\omega t\sigma_z} H(t) e^{-i\omega t\sigma_z} = -2\pi\gamma f(t) B_m D_z(\omega t)\sigma_x D_z(-\omega t), \tag{3.43}$$

where $\omega = 2\pi\gamma B_0$ and D_z is the spin-1/2 axial rotation matrix.

The effect of H' on the system is most readily calculated in a density matrix formalism, as discussed in [76] and [77]

$$\frac{d\rho}{dt} = \Gamma\rho = -\left\langle \int_0^\infty [H'(t), [H'(t-\tau), \rho]] d\tau \right\rangle_{av}, \tag{3.44}$$

where ρ is the 2×2 spin-1/2 density matrix and the average is over a time much longer than τ_c; also assumed is an average over the statistical ensemble represented by the subscript "av". This result comes from the second-order perturbative approximation to the density matrix evolution [see [76], Chap. VIII, equations (28)–(32)]. Γ is referred to as the relaxation matrix.

For simplicity, assume that the particle motion is constant between collisions which on average occur after a time interval τ_c; this results in a particularly simple form for the autocorrelation of $f(t)$:

$$f(t)f(t-\tau) = \begin{cases} 0, & \text{if } \tau > \tau_c; \\ 1-\tau/\tau_c, & \text{if } \tau \le \tau_c. \end{cases} \tag{3.45}$$

Ignoring exponential terms with arguments $\omega(\tau + 2t)$ gives

$$\Gamma_{11} = \Gamma_{22} = -\left\langle \frac{(2\pi\gamma B_m)^2}{2} \frac{1-\cos\omega\tau_c}{\omega^2\tau_c} \right\rangle_{av} \tag{3.46}$$

for the diagonal elements of the relaxation matrix, and for the off-diagonal elements,

$$\Gamma_{12} = \Gamma_{21}^* = -\left\langle \frac{(2\pi\gamma B_m)^2}{2} \left(\frac{1-\cos\omega\tau_c}{\omega^2\tau_c} + \mathrm{i}\,\frac{\omega\tau_c - \sin\omega\tau_c}{\omega^2\tau_c} \right) \right\rangle_{av} . \tag{3.47}$$

The real components of Γ represent the spin relaxation, while the imaginary components of the off-diagonal elements represent a frequency shift; it is

$$\Delta\omega = 2\pi f_m = \frac{1}{2}\left\langle (2\pi\gamma B_m)^2 \frac{\omega\tau_c - \sin\omega\tau_c}{\omega^2\tau_c} \right\rangle_{av} . \tag{3.48}$$

It is interesting to consider the limiting forms of (3.48). When $\omega\tau_c \gg 1$, the term $\sin\omega\tau_c$ has zero ensemble average (given a reasonably broad velocity distribution). Furthermore, taking into account the fact that $\boldsymbol{v}$ is not constrained to lie in a plane perpendicular to $\boldsymbol{E}$, $B_m \to B_m \sin\theta$ must be averaged over all possible directions on a sphere, giving a mean square effect $2B_m^2/3$. Thus, in the limit $\omega\tau_c \gg 1$,

$$f_m = \frac{1}{3}(\gamma B_m)^2/f_0 = \frac{1}{3}(\gamma v E/c)^2/f_0, \tag{3.49}$$

where $f_0 = \gamma B_0$. It should be noted that in this limit the shift does not depend on τ_c, and is just the quadratic sum of the motional and applied magnetic fields. The ultracold neutron storage experiment operates in this regime.

In the case where $\omega\tau_c \ll 1$, the $\sin\omega\tau_c$ term can be expanded

$$f_m = \frac{(2\pi)^2}{9}(\gamma B_m)^2 f_0\tau_c^2 = \frac{(2\pi)^2}{9}(\gamma v E/c)^2 f_0 \tau_c^2, \tag{3.50}$$

where B_m and τ_c represent appropriate ensemble averages. The behavior here is rather unexpected in that the shift increases with f_0, which is opposite to the previous case. Any EDM experiment which employs a buffer gas operates in this regime.

The behavior in the two limiting cases can be qualitatively understood. The time evolution operator for a spin-1/2 system is

$$U = e^{\mathrm{i}Ht} = \cos\gamma|B|t - \mathrm{i}\,\frac{\boldsymbol{\sigma}\cdot\boldsymbol{B}}{|B|}\sin\gamma|B|t, \tag{3.51}$$

and when $\omega\tau_c \gg 1$, the system simply responds to the quadrature sum of all the fields in the problem, as we already knew. The case of $\gamma|B|t \ll 1$ is more subtle. It is useful to work in the rotating frame; the effect of the random field is determined by the magnitude of its static or slowly varying components in that frame. Since the power spectrum of the fluctuations is proportional to the cosine transform of the autocorrelation function, for the rectangular correlation function, the effective random field power is simply the second factor in the average as shown in (3.48). Pictorially, the slowly varying field components lead to a random walk of the spin vector. For small angles, the change in net spin direction is given by the vector sum of all the angular displacements. This gives qualitatively the same answer.

Although we have not yet discussed any EDM experiment in great detail, it is instructive to estimate here the size of possible systematic effects, and the degree to which the electric field magnitude must be controlled to achieve the required level of sensitivity. These experiments will be discussed in greater detail in the appropriate following chapters.

Table 3.1 lists various parameters for existing and proposed EDM measurements. The correlation time is determined through the mean free path L between wall collisions,

$$L = 4V/S, \tag{3.52}$$

where V is the cell volume, and S is the cell surface area; this is a well known result from kinetic theory. The correlation time, for a given average velocity v, is

$$\tau_c = L/v. \tag{3.53}$$

Table 3.1. Parameters and calculated quadratic shifts for various EDM experiments. The electric field E is in MV/m, and $x = 2\pi f\tau_c$

	L [m]	v [m/s]	τ_c [s]	f_0 [Hz]	x	E	γB_m [Hz]	f_m [nHz]	f_e [nHz]	d [e cm]	$\delta E/E$
UCN	0.1	5	0.02	30	3.8	1	0.0016	30	100	10^{-26}	–
^{199}Hg	–	120	10^{-9}	7.5	10^{-7}	1	0.01	10^{-12}	1	10^{-28}	–
UCN–^{199}Hg	0.24	3 120	0.08 0.002	30 7.5	15 0.09	1	0.001 0.01	12 0.2	10	10^{-27}	0.5
UCN–^{3}He	0.05	3 30	0.02 0.002	1 1	0.1 0.01	10	0.01 0.1	180 240	1	10^{-29}	0.01

In Table 3.1, f_e represents the desirable limit of the motional field effect to ensure that, for a given experiment, it does not contribute a systematic error (also listed is the EDM magnitude to which f corresponds). The final column gives the precision with which $\boldsymbol{E}$ must be reversed to achieve this accuracy; since f_m is quadratic in E, the change in f_m is related to a variation in the magnitude of $\boldsymbol{E}$ by

$$f_e = 2f_m \frac{\delta E}{E} = \frac{1}{10} f_{stat}, \tag{3.54}$$

where f_{stat} is the final experimental statistical uncertainty.

The parameters for the completed ultracold neutron (UCN) EDM experiments are listed first; the two most precise results have the same final uncertainty and are quite similar in their experimental parameters [78, 79]. It can be seen that a possible motional field systematic effect is safely below the reported uncertainties.

The ^{199}Hg atomic EDM experiment is listed next [80]. This experiment employs neutral atoms stored in an optical pumping cell along with 200 torr of N_2 buffer gas; hence the correlation time, being determined by collisions with the buffer gas, is very short. This result illustrates the importance of buffer gas as a means to reduce the motional field effects; however, for this experiment, given that the cell mean free path is of order 1 cm, the shift without buffer gas would be only 0.02 nHz.

The UCN$-^{199}$Hg comagnetometer EDM experiment which is just beginning operation is listed next; the idea here is that spin-polarized ^{199}Hg is stored along with the UCN to give a nearly direct measure of the average magnetic field seen by the UCN [63]. The storage cell for this experiment is larger than that used in the earlier experiment and also has a lower wall storage potential, hence a lower average velocity. As shown in Table 3.1, the effect of the motional field on the ^{199}Hg magnetometer is insignificant, while the effect for the UCN is just at the level where one might be concerned. It is interesting to note that the species with the highest velocity, hence the largest B_m, has a considerably reduced effect because of the short correlation time.

The final experiment listed is a proposed UCN$-^3$He comparison, done in a superfluid ^{4}He bath at a temperature 0.5 K [81]. This experimental proposal is fully described in Sect. 4.6. In the light of various technical constraints, the original proposed storage volume has been reduced in size, and the currently proposed storage volume is a 10 cm diameter by 5 cm high cylinder. Also proposed is a substantially increased electric field strength, and Table 3.1 shows that the anticipated shifts are rather large. The listed constraint on $\delta E/E$ is in fact a worst-case estimate, based on a simple comparison of precession frequencies. In arriving at this constraint, an important point was neglected: the experimental technique relies on use of a dressed-spin technique, fully described in Sect. 4.6, to effectively make the gyromagnetic ratios equal, eliminating the effect of static magnetic fields altogether. However, the motional field effect is unlike an applied static magnetic field in that its magnitude depends on the ensemble average properties of a given species. Even so, in the case where $\omega\tau_c \ll 1$, the dressed spin technique tends to reduce the relative motional field effect; if v in (3.50) is rewritten as L/τ_c, so that

$$f_m = \frac{(2\pi)^2}{6}\left(\gamma\frac{L}{\tau_c}E/c^2\right)^2 f_0\tau_c^2 = \gamma^3\frac{(2\pi)^2}{6}(LE/c^2)^2 B_0, \qquad (3.55)$$

and all species-specific properties, other than the gyromagnetic ratio, are eliminated. Furthermore, it is interesting to note that $LE \approx V$, the voltage applied to the system. The main point is that the technique relies on detecting a relative precession between the UCN and ^{3}He when a dressed spin technique is used to set $\gamma_n = \gamma_{^3He}$; the addition of the motional field changes the net static field magnitude and direction. Since the dressing condition varies only quadratically with the angular deviation from perpendicularity of the static and dressing fields, we see that the motional field effect will be substantially reduced in this case. Furthermore, the shift in this limit is proportional to the applied static magnetic field amplitude. This implies operation at the lowest possible magnetic field, and that reversal of its direction relative to the electric field will not change the frequency shift, unlike for a true EDM.

The above results are approximate in that a particularly simple form for the autocorrelation function was used, based on the assumption that the species travel on undisturbed paths between collisions. For UCN, this is not true because at the low velocities of < 5 m/s, the paths are parabolic. Nonetheless, deviations from linear trajectories are expected to give average corrections in the range of a few percent. Furthermore, since the magnitude of the velocity of a particular UCN does not change on collision with the wall, the average of the motional field shift over a long time for the chosen UCN does not give the ensemble average. For an atom, the single-particle average represents the ensemble average, given the high likelihood of a velocity magnitude and direction change during a wall collision. In the case of relatively heavy Hg atoms diffusing in an N_2 buffer gas, several collisions are required before the velocity becomes completely uncorrelated

4. The Search for the Neutron EDM

> There are at least two kinds of games. One could be called finite, the other infinite.
> A finite game is played for the purpose of winning, an infinite game for the purpose of continuing the play.
> *James P. Carse*

The neutron EDM has interested physicists for half a century; indeed, the original suggestion of the possibility of a neutron EDM, as described in Chap. 1, can be considered as the progenitor of all questions of symmetry violation in fundamental interactions, a mainstay of modern physics. Now, some 45 years after the original suggestion, the neutron EDM is perhaps more interesting than ever. This is in part due to the great improvements in experimental sensitivity which have been achieved over the years, and the expectation for future improvements. Figure 4.1 shows the historical development of the upper limit of the neutron EDM, along with some theoretical predictions. It can be seen that the neutron EDM limit has been decreasing, at a nearly constant rate, by about two orders of magnitude per decade. (Such a constant rate of change is well known in, for example, the maximum energy of particle accelerators vs. time; the constant rate is simply a psychological result–an experiment isn't worth doing unless one achieves at least a factor of ten improvement, and funding agencies plan in five-year units of time.) In this chapter, a brief review of neutron physics will be presented, followed by a discussion, in chronological order, of the experiments (or proposals) that lead to Fig. 4.1. Reviews of the neutron EDM experimental search are given in [43, 45, 82]. The earliest limit on the neutron EDM was obtained in 1949 by Ramsey and Purcell, which was inferred from direct slow neutron scattering from the interaction of an EDM with an atomic Coulomb field in Pb [4]. The limit in this case is 3×10^{-18} e cm, and was determined from data reported in 1947 [83, 84].

4.1 Properties of the Neutron

Neutrons account for about half of all the mass around us; yet, the neutron was only discovered in 1932, the same year that the positron was discovered,

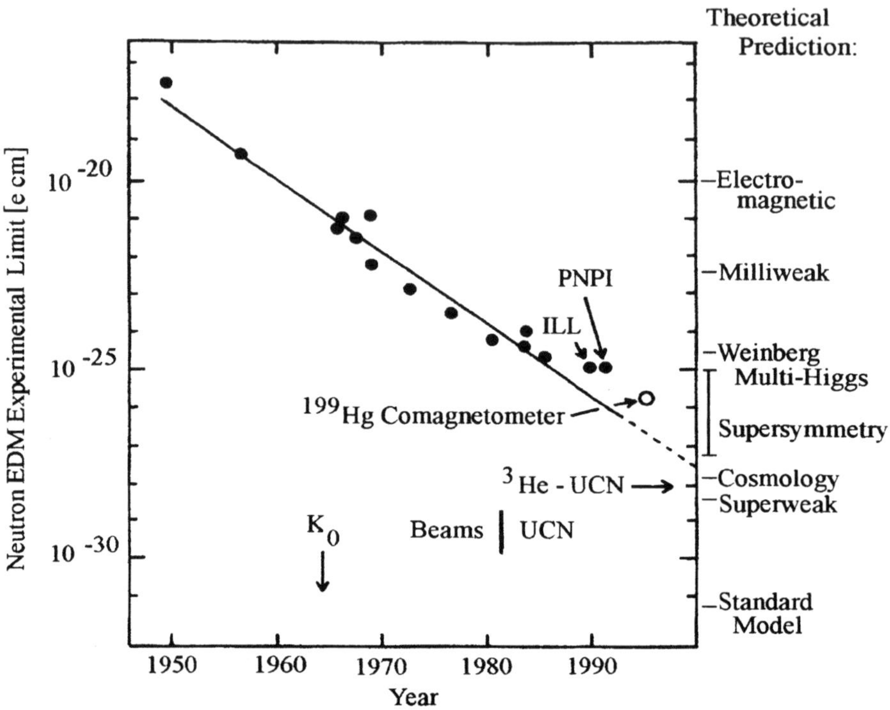

Fig. 4.1. Historical development of the neutron EDM experimental limit along with some theoretical predictions

and the year in which deuterium was isolated for the first time. This late discovery as compared to the electron and proton is due in part to the fact that the free neutron is unstable with respect to β decay, with a lifetime now known to be 887 ± 2 s, with the most accurate values obtained from ultra-cold neutron storage experiments. In fact, the existence of the neutron was postulated as early as 1923 to explain the possible existence of isotopes (see [85] for a discussion of the early work). Originally, the neutron tended to be thought of as a super-bound hydrogen atom; the neutron mass as first measured indicated that indeed this model might be correct; subsequently, it was realized that the gamma energy associated with the photodisintegration of the deuteron allows a very precise neutron mass measurement, and the result indicated that the neutron was probably unstable with respect to β decay [86]. Neutron β decay was first directly observed in 1948 [87]. From atomic hyperfine interactions, it was known that the neutron magnetic moment must be on the same order as that of the proton. From molecular spectroscopy, it was known that diatomic molecules with odd-A nuclei obey Fermi–Dirac statistics. It indicated also that the neutron intrinsic angular momentum must be half-integer.

Measurements of the deuteron magnetic moment, most notably by Rabi and associates, were the first indications that the neutron itself has an intrinsic magnetic moment; in these experiments, it became evident that the deuteron has a quadrupole moment, which represented the discovery of the tensor character of the nuclear forces [51]. Such a discovery is in the spirit of the work described in this book, that is, inference of rather high energy interactions leading to subtle low energy effects. The results obtained by Rabi et al. could not uniquely determine the magnitude of the neutron magnetic moment nor its sign; the first quite definite indication of the neutron spin and magnetic moment magnitude was found by Fermi [88]. It was shown that the observed rate of capture of neutrons by hydrogen (protons) could be well described by a spin–spin interaction assuming both the neutron and proton have spin 1/2, and this can be considered the first firm proof that the neutron is a spin-1/2 system. From the deuteron magnetic moment, it can be inferred that the neutron magnetic moment is negative. This was subsequently directly shown through neutron beam magnetic resonance experiments; the neutron magnetic moment is now known to a few parts in 10^7. Table 4.1 lists the currently accepted properties of the neutron, as given by the Particle Data Group [89].

Table 4.1. Currently accepted properties of the neutron

Property	Value
Intrinsic spin I	1/2
Mass m_n	939.56563±0.00028 MeV
Mean life τ	887.0±2.0 s
Magnetic moment μ	$-1.9130428\pm$ 0.0000005 $\mu_{\rm n}$
Electric dipole moment d	$< 10 \times 10^{-26}$ e cm (95% conf)[a]
Electric polarizability α	$(1.16 \pm 0.21) \times 10^{-3}$ fm^3
Charge q	$(-0.4 \pm 1.1) \times 10^{-21}$ e

[a] The value as presented here will be discussed in Sect. 4.5.1

4.2 Interaction of Neutrons with Matter

The interaction between a neutron and a proton can be represented (when the relative velocity is sufficiently small) by an attractive spherical square-well potential with a depth U_0 of about 40 MeV and a radius R of about 2 fermi (1 fermi= 10^{-15} m). The force between a neutron and a heavier nucleus has essentially the same form, with the well depth remaining nearly constant, and radius growing approximately as $r_0 A^{2/3}$, where A is the total number of nucleons (constant nuclear density). For our present discussion, we can assume that there is no neutron–electron interaction.

Slow neutrons, by definition, have a de Broglie wavelength $\lambda = 2\pi\hbar/mv$ (m is the neutron mass, v the neutron velocity) which satisfies

$$\frac{2\pi R}{\lambda} = kR \ll 1. \tag{4.1}$$

In the region where $r > R$, i.e., where the neutron–nucleus potential $U(r) = 0$, we can write the total neutron wave function (incident plane wave plus scattered wave) as

$$\psi = \mathrm{e}^{\mathrm{i}\boldsymbol{k}\cdot\boldsymbol{r}} + f(\theta)\frac{\mathrm{e}^{\mathrm{i}kr}}{r}. \tag{4.2}$$

When (4.1) is satisfied, the scattering is predominately s wave, and we can write

$$f(\theta) = \text{constant} = b = -a, \tag{4.3}$$

where a is called the coherent scattering length, and b the scattering amplitude. The differential scattering cross-section is

$$\frac{\mathrm{d}\sigma}{\mathrm{d}\Omega} = |f(\theta)|^2 \tag{4.4}$$

which in the case of s wave scattering is constant.

In the range $R < r < 1/k$, equation (4.2) can be written in the form

$$\psi \approx 1 - a/r = (r-a)/r. \tag{4.5}$$

Since $\psi = 0$ at $r = a$, a can be interpreted as the radius of a hard sphere which would produce the same scattered wave function at $r \gg a$ as does the actual potential. When the sign of a is defined as in (4.3), as was done originally in [90], $a > 0$ indicates that the wave function is pushed away from the origin, and one would conclude, on looking at the scattered wave function at great distances from the origin, that the neutron–nucleus interaction is repulsive.

If we transform the scattered wave function to $u = r\psi$, in the region $r < R$, u has the form

$$u = A\sin Kr, \qquad K = \sqrt{k^2 + \frac{2mU_0}{\hbar^2}}, \tag{4.6}$$

which must satisfy the boundary conditions $u = 0$ at $r = 0$ (guaranteed) and that u'/u be continuous at $r = R$. From (4.5) and (4.6),

$$a/R = 1 - \frac{\tan KR}{KR}. \tag{4.7}$$

Only when $\tan KR > KR$ is it possible that $a < 0$; thus, except for $KR < \pi/2$, the likelihood that $a < 0$ is very small. It turns out that $KR > 3$ for all nuclei, with the exception of hydrogen, so nuclei with negative scattering lengths are expected to be rare. This is observed experimentally, with $a < 0$ for hydrogen and only three or four other known isotopes. Since the strong interaction is spin dependent, a can also be spin dependent (see [15], §132).

In addition to being scattered, neutrons can also be captured by nuclei. This process, called absorption, can lead to the emission of a γ ray or charged particles. The probability for such reactions is proportional to $|M|^2$/(incident flux) where

$$M = \langle \psi_\mathrm{f} | V | \psi_\mathrm{i} \rangle. \tag{4.8}$$

Here $\psi_{\mathrm{f,i}}$ refer to the incident and final neutron/nucleus compound states, and V is the neutron–nucleus interaction. By arguments similar to those showing that $f(\theta)$ is independent of incident neutron energy, it can be shown that M is also. Thus, the absorption probability is proportional to

$$1/(\text{incident flux}) = 1/v.$$

A neutron beam of intensity I_0 will be attenuated on propagation through matter by

$$I(L) = I_0 \mathrm{e}^{-\rho\sigma L} \tag{4.9}$$

where ρ is the number of nuclei per unit volume; σ, usually proportional $1/v$, is the absorption cross-section; and L is the distance into the target that the neutron beam has propagated. It is also possible to think of absorption as simply a rate of neutron loss while in the presence of matter:

$$\sigma L = \sigma_0 \frac{v_0}{v} L = \sigma_0 v_0 t, \tag{4.10}$$

where σ_0 is the absorption cross-section measured at v_0, and $t = L/v$. Note that no assumptions have been made about the velocity of the absorbers; if fact, even for neutron absorption by a gas, the relevant velocity is that of the neutrons relative to the center of mass of the absorbers. The relevant parameter is the time that the neutron spends in the vicinity of the absorbers.

To describe the interaction of slow neutrons with bulk matter or molecules, Fermi developed the concepts of the effective or "pseudo" potential and the neutron index of refraction [90] (see also [15], §151.) His idea is as follows. Although the range of nuclear forces is small, they are quite strong within that range so one cannot in general apply perturbation theory to a collision between a neutron and a nucleus. However, as we have shown, the scattering amplitude for slow neutrons is a constant independent of incident neutron velocity. The constant amplitude can be reproduced if we describe the neutron–nucleus interaction as a point interaction

$$U(\boldsymbol{r}) = -\frac{2\pi\hbar^2}{M} f \delta(\boldsymbol{r}), \tag{4.11}$$

where M is the neutron–nucleus reduced mass and f is the constant scattering amplitude. When this potential is substituted into the Born approximation,

$$f = -\frac{M}{2\pi\hbar^2} \int U(\boldsymbol{r}) \mathrm{e}^{-\mathrm{i}\boldsymbol{q}\cdot\boldsymbol{r}} \mathrm{d}V, \tag{4.12}$$

the δ function makes the integral independent of momentum transfer $\boldsymbol{q}$.

When many scatterers are bound in a piece of bulk matter, the potential seen by a neutron is

$$U_F(\boldsymbol{r}) = \frac{2\pi\hbar^2}{m}\sum_i a_i\delta(\boldsymbol{r}-\boldsymbol{r}_i), \tag{4.13}$$

where $m \approx M$, and i labels a particular nucleus bound in the material ($a_i = -f_i$). Thus, a neutron incident on a piece of matter (solid, liquid, or gas) will see a "forest" of delta-function potentials. The neutron wave function will then consist of the incident wave and a sum of the spherical waves scattered by each nucleus, as in (4.2):

$$\psi(\boldsymbol{r}) = \mathrm{e}^{\mathrm{i}\boldsymbol{k}\cdot\boldsymbol{r}} - \sum_i a_i \frac{\mathrm{e}^{\mathrm{i}k|r-r_i|}}{|r-r_i|}\mathrm{e}^{\mathrm{i}\boldsymbol{k}\cdot\boldsymbol{r}_i}. \tag{4.14}$$

This is the usual starting point for calculation of the effects of bulk material on the neutron wave function (see [91] or [92] for a more complete treatment, and [93] for a discussion of appropriate approximations). However, we will take a less rigorous approach here: one which gives some additional physical insight into the problem.

The effect of the potential given by (4.13) on an arbitrary neutron wave packet can be determined by first-order perturbation theory. Again, this approach deviates from the standard treatment. Consider a neutron wave packet inside an extended material body:

$$\psi(\boldsymbol{r}) = \int g(\boldsymbol{k})\mathrm{e}^{-\mathrm{i}\boldsymbol{k}\cdot\boldsymbol{r}}\mathrm{d}^3k. \tag{4.15}$$

To first order, the effect of the material on the neutron total energy is

$$\begin{aligned}\Delta E &= \langle\psi|U_F|\psi\rangle \\ &= \frac{2\pi\hbar^2}{m}\int\!\!\int\!\!\int \mathrm{e}^{\mathrm{i}\boldsymbol{k}\cdot\boldsymbol{r}}\mathrm{e}^{-\mathrm{i}\boldsymbol{k}'\cdot\boldsymbol{r}}g^*(\boldsymbol{k})g(\boldsymbol{k}')\sum_i a_i\delta(\boldsymbol{r}-\boldsymbol{r}_i)\mathrm{d}^3r\mathrm{d}^3k\mathrm{d}^3k'.\end{aligned} \tag{4.16}$$

If the bound scatterers are on a regular array, then the sum of δ functions can be expanded in a spatial Fourier series in $\mathrm{e}^{\mathrm{i}\boldsymbol{k}''\cdot\boldsymbol{r}}$, and the integral is non-zero when

$$-\boldsymbol{k}+\boldsymbol{k}' = \boldsymbol{k}'' \tag{4.17}$$

where $\boldsymbol{k}''$ represents a reciprocal lattice vector; this is diffraction. We are interested in the long wavelength limit, and elastic scattering: $|\boldsymbol{k}| = |\boldsymbol{k}'| \ll |\boldsymbol{k}''|$. On the other hand, for the case of a perfectly homogeneous material, with no internal periodicity in the spatial distribution of atoms, $\boldsymbol{k}'' = 0$. (This can be thought of as the long neutron wavelength limit because in this limit there is no atomic periodicity commensurate with the neutron wavelength, i.e., the neutron energy is below the Bragg cutoff.) We can therefore take $\boldsymbol{k}'' = 0$ in (4.17) (forward scattering only). Then we know that contributions to the neutron energy come only in first order and there is no mixing between different $\boldsymbol{k}$; the corrections to the energy can be written

$$\Delta E = \frac{2\pi\hbar^2}{m} \int |\psi(\boldsymbol{r})|^2 \sum_i a_i \delta(\boldsymbol{r} - \boldsymbol{r}_i) \mathrm{d}^3 r. \tag{4.18}$$

We can choose the neutron wave function to be of arbitrary shape; assume it is constant through a small spherical volume V. Then (4.18) becomes

$$\Delta E = \frac{2\pi\hbar^2}{m} \sum [\rho a]_k / V = \frac{2\pi\hbar^2}{m} \langle \rho a \rangle_V, \tag{4.19}$$

where $[\rho a]_k$ is the number density and scattering length for a particular isotope, and $\langle\rangle_\mathrm{V}$ represents an average over V of the scattering lengths and densities of the different isotopes within the small volume. If we consider a neutron with total energy $E = \hbar^2 k_0^2/2m$ outside the material, inside the total energy is the same, so

$$k_0^2 = k^2 + 4\pi \langle \rho a \rangle_V. \tag{4.20}$$

In fact, there is no restriction on V since the wave packet can be made arbitrarily small; we can therefore consider a larger piece of material made of smaller subunits, each with $\langle \rho a \rangle_V = [\rho a(\boldsymbol{r})]$ which represents a possible spatial density and composition variation over the larger piece of material.

Since k changes on entering the material, we can define a material index of refraction

$$n(\boldsymbol{r}) = \sqrt{1 - \frac{4\pi [\rho a(\boldsymbol{r})]}{k_0^2}}, \tag{4.21}$$

or an effective potential

$$U(\boldsymbol{r}) = \frac{2\pi\hbar^2}{m} [\rho a(\boldsymbol{r})]. \tag{4.22}$$

For a pure homogeneous material,

$$U_F = \frac{2\pi\hbar^2}{m} \rho a. \tag{4.23}$$

Thus, the surface of a .piece of bulk matter represents a potential step of height U_F to an entering neutron. Most interestingly, this potential is *repulsive* for the vast majority of nuclei ($a > 0$) even though the neutron–nucleus potential is attractive. This is because the effective potential is the volume average of the Fermi potential, which is repulsive for an attractive neutron–nucleus potential; this is a consequence of matching the boundary conditions at the nucleus surface [see (4.5)]. Since the potential U_F is independent of neutron energy while the neutron index (4.21) depends on the neutron energy, for many applications, the potential is the more useful quantity.

Absorption can be modeled as an imaginary potential; we saw before that the rate of neutron absorption within a material is given by $\rho\sigma_0 = \Gamma$. If we write

$$\psi(t, \boldsymbol{r}) = \mathrm{e}^{(-\mathrm{i}\omega - \Gamma/2)t}\psi(\boldsymbol{r}), \tag{4.24}$$

where ω represents the total energy, we see the neutron probability decreases as $\mathrm{e}^{-\Gamma t}$. Since the neutron wave function, if it is an eigenstate, evolves as $\mathrm{e}^{-\mathrm{i}\Lambda t}$ where Λ is the eigenvalue, it is apparent that

$$\Lambda = \frac{\hbar^2}{2m}k_0^2 + U_F + \mathrm{i}\Gamma/2 \tag{4.25}$$

for a neutron with wavenumber k_0 outside the material. This motivates us to write the Hamiltonian as

$$H = p^2/2m + U_F - \mathrm{i}W, \tag{4.26}$$

where $W = \hbar\rho\sigma_0 v_0/2$.

Equation (4.26) forms the basis of the field of neutron optics (see, for example, [91]), and represents the one dimensional potential step well known from all introductory texts on quantum mechanics, except here we have allowed for a complex potential to describe the absorption of neutrons.

If a neutron is incident obliquely on a material surface at an angle θ relative to the surface normal, we can transform into a frame where the component of the velocity parallel to the surface is zero; if the kinetic energy given by the perpendicular component of the velocity is less than U_F for the material, the neutron will be totally reflected from the surface. Table 4.2 lists the properties and effective potential of some elements and common materials.

This can also be understood from an optical point of view. The refraction of neutrons at material surfaces is exactly analogous to that of light waves (for the electric vector perpendicular to the plane of incidence; in this case, the boundary conditions are identical) since both light and neutrons satisfy the same wave equation. Thus, on entering a material ($n < 1$) from vacuum ($n = 1$), neutrons will be "totally externally reflected" for glancing incidence angles (measured from the surface) less than the critical angle

$$\theta_c \approx [2(1-n)]^{1/2} \approx \lambda(\rho a/\pi)^{1/2} \tag{4.27}$$

which depends on the neutron energy.

Equation (4.21) can be rewritten

$$n = (1 - U_F/E)^{1/2}, \tag{4.28}$$

where E is the incident neutron kinetic energy. The effective potential U_F is between 100 and 300 neV (1 neV= 10^{-9} eV) for most materials, giving $\theta_c \approx 17$ mrad $\approx 1°$ for 2 meV (20 K thermal energy) neutrons.

Table 4.2. Neutron optical properties of selected materials

Element	ρ $[10^{22}\text{cm}^{-3}]$	$\langle a_{\text{coh}}^{\text{bound}} \rangle$ [fm]	U_F [neV]	v_{crit} [m/s]	$\sigma_0(18\text{Å})$ [barn]	$f = W/U_F$ $[\times 10^{-5}]$
^{58}Ni	9.0	14.4	335	8.0	44	8.6
C (diamond)	17.6	6.6	303	7.6	-	-
BeO	7.25	13.6	261	7.1	6.6	1.35
Ni	9.0	10.6	252	7.0	48	12.5
Be (300 K)	12.3	7.75	252	7.0	1.4	0.5
^{65}Cu	8.93	11.0	224	6.6	28	7.0
Fe	8.5	9.7	210	6.3	30	8.5
C (graphite)	10.0	6.6	180	5.9	30	8.5
D-Plexiglas $(C_2D_3O)_n$	1.65	39.11	168	5.7	-	-
D-Polystyrene $(CD)_n$	4.76	13.3	165	5.6	-	-
Teflon $(CF_2)_n$	2.65	17.6	123	4.9	-	-
Pb	3.29	9.6	83	3.	2.0	0.6
Al	6.02	3.45	54	2.0	2.8	2.25
Plexiglas $(C_2H_3O)_n$	1.65	7.88	33.9	2.6	1.0	
V	7.1	−0.382	−7.2	-	50	-
Polyethylene $(CH_2)_n$	3.9	−0.84	−8.7	-	-	
H_2O	3.34	−1.68	−14.7	-	-	
Ti	5.6	−3.34	−48	-	58	-

4.2.1 Neutron Polarization

A neutron polarizer can be made by applying a thin magnetized ferromagnetic film to a highly absorbing surface. Neutrons in the appropriate spin state will, through the Zeeman interaction $\boldsymbol{\mu} \cdot \boldsymbol{B}$, have a larger critical angle and be reflected while those in the other state will be transmitted by the film and lost in the absorbing surface. Such polarizers have become quite sophisticated and yield beam polarizations of better than 95%. Development of such mirror polarizers was a significant achievement; previously, neutron polarization was based on small angle scattering in bulk magnetized iron. The transmission and polarization by this technique was at best marginal; Ramsey has some commentary on the subject [51].

With modest guiding magnetic fields, the neutron spin polarization for a neutron beam can be easily maintained if the neutron motion is so slow that the magnetic moment always keeps the same orientation with respect to the magnetic field $\boldsymbol{B}$ (adiabatic case). The necessary condition is that the time dependence of the field, as seen by the neutron moving through the inhomogeneous field, is much less than the precession frequency of the neutron moment in the field (taking τ as the characteristic time scale of the field change):

$$\frac{1}{\tau} = \frac{1}{B}\left|\frac{\mathrm{d}\boldsymbol{B}}{\mathrm{d}t}\right| \ll \frac{\boldsymbol{\mu}\cdot\boldsymbol{B}}{\hbar} = \omega_{\mathrm{L}}. \tag{4.29}$$

It is usually expected that, in this limit, the polarization loss is exponentially small (see [15], §53).

Under actual conditions encountered in the laboratory, there are usually rapidly varying (in the neutron frame) fields that can drive spin-flip transitions. Consider a simple example where the tip of a constant magnitude magnetic field B_0 sweeps out an arc of angle α over a time τ. In this case, a transformation to a rotating frame is perfectly valid, with no approximations. In this frame, in addition to the now static field B_0, there is a field along the rotation axis, perpendicular to B_0, of magnitude

$$B_\perp = \frac{1}{\gamma}\frac{\alpha}{\tau}.$$

If the neutron spin was originally along B_0 when the field direction started to change, then in the rotating frame the neutron precesses about a new axis, the vector sum of B_0 and $B_\perp$. If the neutron makes many precessions during the time τ, and if there are either slight gradients in the field or a finite width velocity distribution, the neutron spin (average expectation value) is projected onto this new axis. The angle between the vector sum and the original neutron spin direction is $\theta \approx B_\perp/B_0$. Therefore, on transforming to the rotating frame, the spin polarization is reduced by a factor $\cos\theta$; this factor applies again on transforming back to the laboratory frame. The net spin flip probability is

$$1 - \cos^2\theta \approx \left(\frac{B_\perp}{B_0}\right)^2 = \frac{\Omega^2}{B_0^2\gamma^2},$$

where $\Omega = \alpha/\tau$ is the rate of change of B_0. This is an essentially exact calculation for the case considered, and is certainly not exponentially small; a practical limit for the spin flip (transition) probability is of order $1/(\omega_{\mathrm{L}}\tau)^2$. Passage through a region of zero magnetic field usually results in complete loss of neutron polarization, as the above implies.

The above case actually illustrates a failure of adiabaticity; however, when neutron losses are controlled to the level of $1/(\omega_{\mathrm{L}}\tau)^2$, experimentalists usually say that the spin transport is adiabatic. The loss of adiabaticity in this example occurs when the tip of the magnetic field begins to move, which is taken as instantaneous.

When the adiabatic condition is badly violated ($\omega_{\mathrm{L}}\tau \ll 1$), the neutron dipole moment tends to keep its direction fixed in space (the sudden approximation), but only for a short time if the fields are large.

An inhomogeneous magnetic field will exert a force on a neutron,

$$\boldsymbol{F}_{\mathrm{m}} = -\boldsymbol{\nabla}\boldsymbol{\mu}\cdot\boldsymbol{B} = \pm|\boldsymbol{\mu}|\boldsymbol{\nabla}|\boldsymbol{B}(\boldsymbol{r})|, \tag{4.30}$$

where, in the last equality, it is assumed that the adiabatic condition applies.

4.2.2 Production and Moderation of Neutrons

The earliest neutron experiments made use of radium–beryllium or radium–polonium sources, with neutrons being released in the reaction

$$\alpha + {}^{9}\mathrm{Be} \rightarrow {}^{12}\mathrm{O} + \mathrm{n}.$$

Cyclotrons can also be used as the α source, as was done in some early neutron magnetic moment experiments. At present, the most intense sources are nuclear reactors, with maximum available thermal (i.e., moderated to 300 K) neutron fluxes of order 10^{15} n/(cm^2 s). Spallation sources, based on the bombardment of a high atomic weight target by an intense GeV proton beam, might become competitive with reactors in the near future. This point will be further discussed later in relation to a possible future neutron EDM experiment.

The neutrons initially released in such reactions have energies in the range of a few MeV for reactors to about 100 MeV for spallation sources. In order for the neutron optics arguments presented earlier to be valid, the neutron wavelength must be at least a few Å (interatomic spacing in solids), corresponding to energies less than 0.03 eV (= 30 meV). The fast neutron energy can be dissipated by scattering from light nuclei; this process was quite accidentally discovered by Fermi and coworkers in 1934 (see [94] for a detailed and anecdotal discussion of this discovery).

The basic idea of moderation is that if a liquid or solid material (a moderator) is placed close to the core of a nuclear reactor, and if the nuclei do not absorb neutrons and the material is of infinite extent, energetic neutrons will eventually come into thermal equilibrium with the moderator, acquiring a Maxwellian energy distribution with temperature given by that of the moderator. This is because when an energetic neutron collides with a proton at rest, the initial kinetic energy is divided equally between the neutron and proton. Thus, after 10 to 20 scatterings, the neutrons released by fission or spallation approach zero energy, with a lower bound corresponding to the temperature of the moderator. Nuclei with higher mass are less effective in direct moderation, but are still useful, heavy water and graphite being two examples. In any real system, the moderator is of finite extent, and also absorbs neutrons according to $1/v$, thereby distorting the Maxwellian spectrum; in most cases, however, it is useful to approximate the neutron spectrum as Maxwellian with a temperature slightly higher than that of the moderator.

The thermal equilibrium density of neutrons with velocities between v and $v + dv$ is given by

$$\rho(v)\mathrm{d}v = \frac{2\Phi_0}{\alpha}\frac{v^2}{\alpha^2}\exp(-v^2/\alpha^2)\frac{\mathrm{d}v}{\alpha} \rightarrow \int v\rho(v)\mathrm{d}v = \Phi_0, \tag{4.31}$$

where Φ_0 is the total thermal flux which characterizes the strength of the neutron source, $\alpha = (2k_\mathrm{B}T_\mathrm{n}/m)^{1/2}$, and T_n is the neutron temperature.

There is a lower practical limit to the temperature that a neutron can achieve in moderation, regardless of how cold the moderator is. When the neutron energy becomes less than the vibrational energy levels in a solid (or liquid) the effective mass (in regard to neutron collisions) of the atoms that the moderator comprises becomes infinite, leading to severe reduction in the rate of moderation. The effective neutron temperature of a liquid deuterium or hydrogen moderator maintained at 20 K is about 25 to 30 K. Other materials, such as solid methane, achieve similar neutron temperatures. Neutrons of this temperature, corresponding to a wavelength of about 6 Å or a kinetic energy of 3 meV, are usually referred to as cold neutrons [compared to 1.8 Å for "thermal" (room temperature energy) neutrons].

For condensed matter studies, neutrons with just that energy are among the most interesting for inelastic scattering and structure measurements. For fundamental work such as EDM measurements, one would like the coldest possible neutrons to maximize the interaction time, for example, in a beam experiment. Fortunately, the cold neutron spectrum is close to a true thermal distribution, so there is in fact a substantial density of neutrons at lower energies. The art of neutron EDM experiments is in making efficient use of the lowest possible energy neutrons.

4.2.3 Transport of Cold Neutrons

Neutron guides, which are essentially evacuated pipes with the inner surface coated with a very smooth layer of high effective potential material, operate on the principle of total internal reflection [95]. The neutrons which diffuse from, for example, a 20 K liquid deuterium moderator, have their velocities isotropically distributed. If the entrance to a neutron guide is placed as close as possible to the moderator, then neutrons with velocities directed within a solid angle determined by the critical angle can be transmitted with very little loss over distances of order tens of meters. Thus, guides provide a source of neutrons some distance from the moderator, but with a flux (within the solid angle) equal to that at the moderator surface. However, the primary advantage of a neutron guide is that is can be slightly bent; faster neutrons and γ rays directly from the reactor core pass through the walls of the guide, and can thus be separated from the experimentally useful slower neutrons which follow the guide in spite of its curvature. Such a separation reduces the experimental background and makes low-counting-rate cold-neutron experimental techniques possible. In fact, the success of the Institut Laue-Langevin in Grenoble, France, is due in large part to such careful shielding; here, the bent guides actually pass through a pool of water several meters long, the water serving as a nearly perfect shielding material.

4.3 Neutron Beam EDM Experiments

4.3.1 The Oak Ridge Experiment of 1950

The first neutron EDM experiment was begun at the Oak Ridge reactor in the spring of 1949 [6]. Although the experiment was completed in 1951, the results were not published until 1957. Figure 4.2 is a rough schematic of the apparatus, and in fact is representative of all beam experiments.

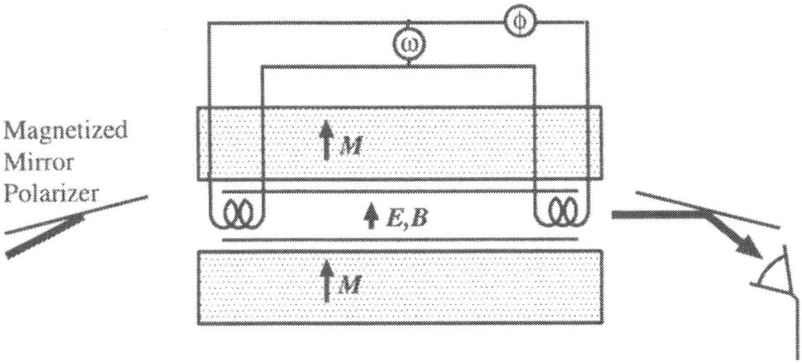

Fig. 4.2. Schematic of a neutron beam EDM apparatus

This experiment was at the forefront of neutron technology. There were four major changes from the earlier magnetic moment experiment [96]. First, the Oak Ridge reactor was used for the source of neutrons rather than a cyclotron target. Second, Ramsey's method of separated oscillatory field was used for measuring the transition frequency (see Sect. 3.1.2). This resulted in much narrower resonances, and the clear region between the separated oscillatory field plates was used to superimpose an electric field, the third difference. The final difference was that a magnetized iron mirror neutron polarizer was used in the EDM experiment.

Up to this time, neutron polarizers and analyzers were based on the bulk scattering in small saturated iron blocks placed between the poles of a powerful electromagnet. The basic idea is that there is a neutron scattering amplitude a for the iron nucleus and a magnetic amplitude $\pm b$ for the neutron dipole–atomic electron dipole magnetic interaction, with the $\pm$ depending on the relative orientation of the spins. The cross section is then $4\pi(a^2+b^2\pm 2ab)$. If all the iron atoms are aligned (saturated sample) the cross sections for the two neutron spin states will differ, and a beam passing through the sample will become polarized. This technique was used in the EDM experiment initially for both the polarizer and analyzer, but better results were later obtained by using a magnetic mirror for the polarizer, a new technique specially implemented in this experiment (see [51] for a discussion of the development).

The homogeneous magnetic field was generated using cold rolled steel pole faces, 180 cm long, 13 cm high, 2.5 cm thick, spaced by 4.4 cm. The pole faces were magnetized by contact with 16 Alnico II magnets bolted uniformly along the bottom of the pole faces. Although the magnetic field was initially quite inhomogeneous, by carefully manually changing the magnetization along the pole faces by exposure to a small magnet or by installation of small iron shims, a homogeneity of about 0.3 G along the neutron flight path was achieved. The static field had magnitude of 250 G, corresponding to a neutron precession frequency of 750 kHz. The separated oscillatory field method does not require a field of high homogeneity. However, the field showed marked temperature dependence and the apparatus was enclosed in a temperature regulated box, holding the temperature constant to within 0.2°. Large steel objects and electromagnets in the vicinity of the apparatus disturbed the field noticeably, but such disturbances are not too troublesome; a differential shift in frequency with reversal of the electric field is the experimental result of interest, and this tends to discriminate random field changes.

The electric field was produced by applying a voltage of 25 kV between two nickel-plated copper plates of length 135 cm, separated by 0.349 cm. This structure was enclosed in a vacuum chamber for insulation, and suspended between the pole faces. The nickel disturbed the field to some extent, but unfortunately the electric field which could be obtained with the bare copper before excessive sparking was considerably lower.

The oscillating magnetic field was produced by two helical coils 5 cm long, spaced 159 cm, with their axes along the neutron beam. The current was supplied by a Hartley vacuum tube oscillator, with stability at the level of 2 Hz, lower than the neutron counting statistical error.

The neutron beam, formed by carefully collimating the direct moderated pile flux, was of dimension 1 cm high by 0.1 cm wide. Unfortunately, the operating conditions of the reactor were secret, and there is no data on the absolute neutron flux [6]. The neutron detector was a BF_3 proportional counter. A thin BF_3 counter with low absorption was placed in the beam to monitor directly the beam intensity as it left the pile to monitor and reject reactor power fluctuations.

Examples of data obtained with this apparatus is shown in Fig. 4.3. The upper curve, with smaller polarizer incident angle, corresponds to neutrons with higher maximum velocity and hence, a wider range of neutron velocities. The higher average velocity is evident in the increased width of the fringes as compared to the lower curve, indicating a shorter time in the interaction region. The increased velocity width is indicated by the more rapid loss of fringe detail in the upper figure.

The solid curves in Fig. 4.3 are theoretical curves with the resonant frequency and peak-to-valley ratio as adjustable parameters. A Maxwellian spectrum of 500 K was used for the neutrons. Again, the second resonance curve at larger mirror angle shows the expected narrowing of the resonance curve

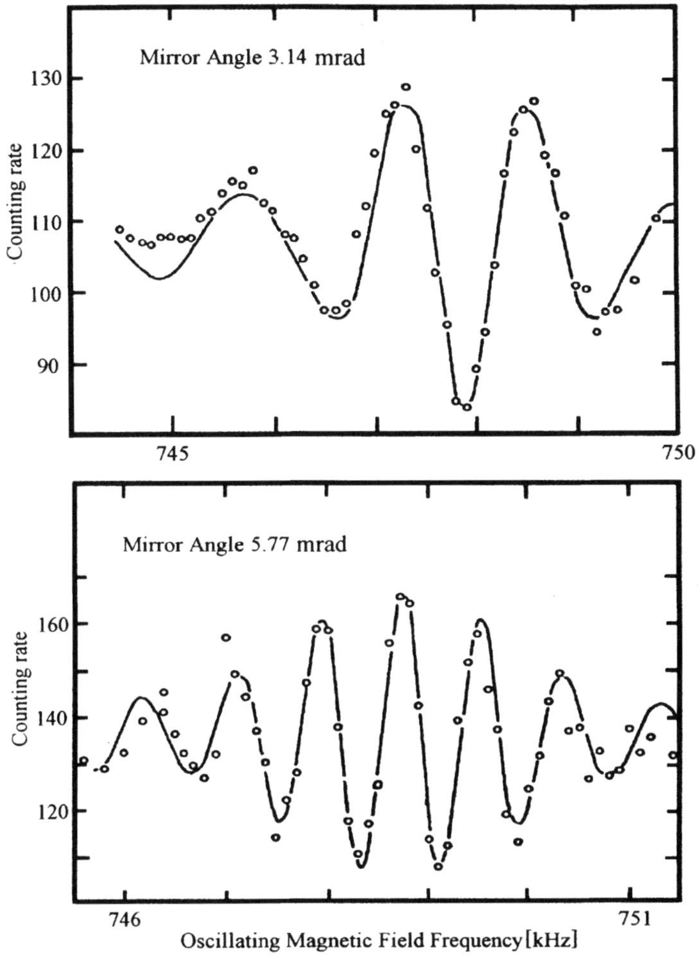

Fig. 4.3. Resonance curves of the neutron counting rate vs. frequency of the oscillating magnetic field from the first neutron EDM experiment

because only the slower neutrons are reflected. The agreement between theory and experiment is excellent; the deviations are within the expected counting statistics. The upper curve corresponds more closely to the parameters used in the experiment. The central peak is a minimum in the upper curve and a maximum in the lower curve because the phase of one of the coils was reversed.

To find the effect of the electric field on the resonance frequency of the neutrons, the oscillating field frequency was set to a point on one of the steep sides of the central resonance. The counting rate as a function of the electric field polarity was then recorded. There is no record of the duration of the counting periods, again because this would allow the reactor flux and power, hence the rate of plutonium production or whatever, to be estimated. However, the total number of counts was around 4 million. Interspersed with the electric field induced counting rate changes, were measurements of the slope at the operating point, which provided a calibration.

The final result, limited by counting statistics, was reported as

$$d_{\mathrm{n}} = (-0.1 \pm 2.4) \times 10^{-20} \; e\,\mathrm{cm}$$

corresponding to an upper limit of

$$d_{\mathrm{n}} < 5 \times 10^{-20} \; e\,\mathrm{cm}.$$

Attempts to measure a systematic magnetic field change correlated with reversal of the electric field polarity were unsuccessful and limited such an effect to a level 10 times below the quoted statistical uncertainty. In this experiment, the average neutron velocity corresponding to the 500 K Maxwellian distribution was 3500 m/s, corresponding to a spin precession coherence time of 0.5 ms.

4.3.2 The Oak Ridge Experiment of 1967

This experiment [97] was performed with the low-intensity test reactor at the Oak Ridge National Laboratory which provided a thermal flux at the core of 3×10^{13} n/(cm^2 s). A small D_2O moderator at the end of a direct core beam hole provided an additionally thermalized flux which illuminated the entrance of a bent neutron guide tube. The part of the neutron tube which was inside the reactor shield was constructed from polished nickel plates and had a 32 m radius of curvature. As discussed above, this curvature allows separation of the relatively slow neutrons from the fast neutrons and gamma rays. The entire neutron path was evacuated to 10^{-4} torr to minimize slow neutron absorption, and to maximize the attainable electric field in the resonance apparatus. A short section of neutron guide with a 1 m radius of curvature, with length about 25 cm, was placed between the shield and the experiment, and served as another low-pass filter on the beam, and reduced the average velocity by a factor of two. The guide dimensions were 10 cm high by 1 cm wide.

Other improvements over the previous experiment included use of 94% Co, 6% Fe magnetized mirror analyzer and polarizers, yielding a polarization of about 55%. The separated field region was 61 cm, and for the 60 m/s average velocity neutrons, gives an interaction time of about 10 ms, or twenty times that of the previous experiment. The electric field was 140 kV/cm, and the magnetic field was 9 G, giving a resonance frequency of 27 kHz. Other important improvements were in the use of magnetic shielding and a ^{6}Li loaded glass scintillator neutron detector. One further difference is that a 90° phase shift between the two oscillating fields was introduced; this results in a "dispersive" resonance pattern, which gives a linear change in counting rate with Larmor frequency at resonance. In the previous beam experiment, the oscillating field was set slightly off resonance in order to obtain a linear sensitivity to a shift in resonance frequency.

The experiment was operated in a similar fashion to the previous one, with the electric field direction reversed every 5 min. The neutron intensity was normalized to a monitor counter to eliminate reactor power fluctuations.

Systematic effects which could be generated by the motional magnetic field (Sect. 3.5.3) were investigated by operating the experiment with the 1 m curvature guide removed, doubling the neutron velocity. No effect was observed, setting a limit on the angle between the electric and magnetic fields to less than 1.5°.

Observations were made with 2000 electric field reversals comprising the counting of 2×10^9 neutrons. The final result was reported as

$$d_{\mathrm{n}} = (-2 \pm 3) \times 10^{-22}\ e\,\mathrm{cm},$$

corresponding to an upper limit of

$$d_{\mathrm{n}} < 6 \times 10^{-22}\ e\,\mathrm{cm}.$$

A slight systematic effect depending on the relative phase of the separated oscillatory fields was observed. To account for this systematic variation, the origin of which was not discovered, the final uncertainty is a factor of two larger than that expected for counting statistics.

4.3.3 The Crystal Scattering Experiment of 1967

This experiment was performed at the MIT reactor and reported in 1967 in the same issue of Physical Review Letters as the previously described experiment [98]. This experiment is based on the interaction of a possible neutron EDM with the Coulomb field in passing through a scattering atom. The atomic electric field had already been shown to scatter neutrons due to the interaction between the $\boldsymbol{v} \times \boldsymbol{E}$ field and the neutron magnetic moment [99]. The need to distinguish between these two effects is a variant of the $\boldsymbol{v} \times \boldsymbol{E}$ problem, discussed in Sect. 3.5.3. (Scattering from the motional field is referred to as Schwinger scattering.)

Figure 4.4 is a schematic representation of the contributions to the scattering by a single atom. The contributions from a possible EDM can be written $V_{\mathrm{e}}(\boldsymbol{r}) = -\boldsymbol{d}_{\mathrm{n}} \cdot \boldsymbol{E}(\boldsymbol{r})$ and the scattering amplitude b'' for this interaction can be calculated in the Born approximation as (see [15], §132)

$$b'' = -\frac{m}{2\pi\hbar^2}\int \mathrm{e}^{\mathrm{i}\boldsymbol{k}_{\mathrm{f}}\cdot\boldsymbol{r}} V_{\mathrm{e}}(\boldsymbol{r})\mathrm{e}^{\mathrm{i}\boldsymbol{k}_{\mathrm{i}}\cdot\boldsymbol{r}}\mathrm{d}\boldsymbol{r} \tag{4.32}$$

where m is the neutron mass, $\boldsymbol{k}_{\mathrm{i,f}}$ are the neutron wavevectors before and after scattering. The electric field has the form

$$\boldsymbol{E}(\boldsymbol{r}) = \frac{Ze}{r^3}\boldsymbol{r}, \tag{4.33}$$

and is modified in the outer regions of the atom by electron screening.

This integral is easily evaluated, yielding

$$-ib'' = \frac{Ze(1-f)d_n \cos\alpha}{\hbar v \sin\theta} \tag{4.34}$$

where $(1-f)$ describes the electron screening, α is the angle between the neutron spin and the direction of the momentum transfer, and 2θ is the angle between the incoming and outgoing wavevectors.

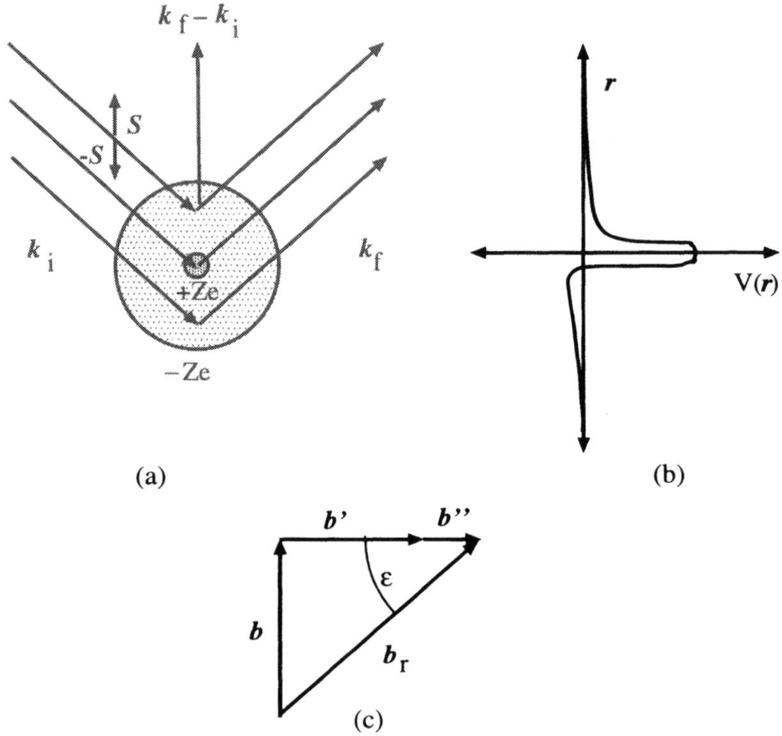

Fig. 4.4. (a) A schematic of neutron scattering from an atom showing the contribution from the nuclear potential and from the interaction between an EDM and the Coulomb field. (b) The volume-weighted potential for the spin direction shown in (a). (c) Vector sum of the real and imaginary components of the scattering length. Reversal of the neutron spin changes the sign of b''

It is interesting to note that the scattering amplitude b'' is imaginary; real or imaginary in this context simply refers to the phase of the scattered wave. As discussed in Sect. 4.2 [see in particular equation (4.26)], the absorption of neutrons by a nucleus can be modeled as an imaginary scattering amplitude; the real part of the scattering amplitude is of order 10^{-12} cm for typical nuclei, and the imaginary part is of order 10^{-16} cm if the absorption cross section is about 1 barn. Some nuclei such as Cd have comparable real and imaginary scattering amplitudes.

The $\cos\alpha$ factor in (4.34) indicates that b'' changes sign when the neutron spin is reversed relative to the momentum transfer. This sign change is equivalent to a phase change of π in the scattered wave which can be converted into an amplitude change. This leads to an intensity change through interference with the nuclear amplitude b. From the limit on the neutron EDM at the time, it was expected that $b''/b < 10^{-4}$. The vector diagram shows that for the EDM scattering b'' to affect the resultant amplitude b_r in first order, $b''+b'$ should be of the same order as b. For CdS, the net scattering amplitude is, for the (004) reflection where the crystal planes alternate between Cd and S (so that the waves reflected from adjacent planes are π out of phase),

$$\begin{aligned} b + \mathrm{i}b' &= b_{\mathrm{Cd}} - b_{\mathrm{S}} \\ &= (0.487 - 0.07\mathrm{i} - 0.285) \times 10^{-12}\ \mathrm{cm} \approx (0.2 - 0.07\mathrm{i}) \times 10^{-12}\ \mathrm{cm}, \end{aligned} \tag{4.35}$$

where b and b' are real numbers; we see a fortuitous cancelation between the Cd and S real scattering amplitudes, and the condition to be linearly sensitive to a neutron EDM is satisfied.

In the actual experiment, Bragg scattering from a CdS crystal was used. In this case, as the incoming wave passes successive planes in the crystal, each makes a contribution to the scattered wave in proportion to the number of atoms in the plane and to the scattering amplitude b_r; when the path length difference for reflections from adjacent planes is one wavelength, the reflections are all in phase and Bragg scattering results. From the theory of Bragg reflection [100] the distance L the neutron wave travels into the crystal before being substantially attenuated by the reflection process is

$$L = \frac{1}{\rho\lambda|b_r|}, \tag{4.36}$$

where ρ is the molecular density and λ the neutron wavelength. Thus, the penetration depth into the crystal depends on the magnitude of b_r, hence on the possible neutron EDM interaction. The intensity of the reflected beam depends strongly on the penetration depth into the absorbing medium; a neutron EDM would be evident by a change in reflected intensity correlated with the neutron spin direction relative to the momentum transfer.

The penetration length L is about 0.4 mm and the decay length due to neutron absorption for Cd is 0.36 mm, so we can assume these two lengths are about equal. In this case, if the uncertainty is due to counting statistics, the statistical limit for an EDM measurement is

$$\sigma(d_n) = \frac{|b_r|\hbar v \sin\theta}{2Ze(1-f)\cos\varepsilon\sqrt{N}}, \tag{4.37}$$

where N is the total number of neutrons counted and ε is shown in the Fig. 4.4c. Using the definition of L and that $\rho \propto (2R)^{-3}$ and $\lambda = 2R\sin\theta$

where R is the distance between scattering planes, the above equation can be rewritten as

$$\sigma(d_n) = \left(\frac{Ze(1-f)\cos\varepsilon}{8R^2}\right)^{-1}\left(\frac{2L}{v}\right)^{-1}\left(\frac{2\sqrt{N}}{\hbar}\right)^{-1}, \tag{4.38}$$

and comparing with (3.5), we can view the first term as the effective electric field, of order 10^9 V/cm, and the second term as the interaction time, of order 10^{-7} s.

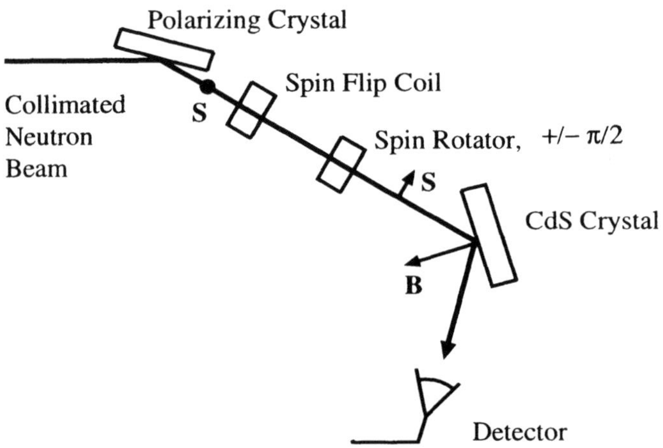

Fig. 4.5. Schematic of the CdS crystal neutron EDM scattering experiment

The schematic diagram of the experiment is shown in Fig. 4.5. A polarized beam of 1-Å neutrons was produced by Bragg scattering from a magnetized cobalt crystal. The resulting beam was highly polarized and very well collimated, hence was perfect for the experiment. A magnetic resonance spin flip coil was placed between the polarizing crystal and the CdS crystal which allowed reversing the incident neutron polarization without disturbing the trajectory. The spin flip coil was followed by an adiabatic spin transport field so that the neutron spin, as it entered the crystal, would be normal to the CdS Bragg planes. The CdS crystal was carefully mounted so that the Bragg planes were normal to the magnetic field which defined the neutron spin direction in the crystal; any deviation from perpendicular would result in a motional field systematic.

By periodically reversing the neutron polarization and by less frequently reversing the specimen crystal and its magnet, it was possible to separate out the motional field effect. Over a three month period 4×10^8 neutrons were counted, and the final result was

$$d_n = (2.4 \pm 3.9) \times 10^{-22}\ e\,\mathrm{cm}$$

corresponding to an upper limit of

$$d_n < 8 \times 10^{-22}\ e\,\mathrm{cm}$$

which is comparable to the previously described beam experiment.

The measured motional field effect was $3.2 \pm 2.0 \times 10^{-22}\ e\,\mathrm{cm}$, corresponding to an angle of 1.6 ± 1.0 mrad between the effective $\boldsymbol{E}$ and applied $\boldsymbol{B}$ fields. The final experimental sensitivity was limited by neutron counting statistics. Given the strict requirement on angular adjustment and the limited choice of scattering crystals, it is unlikely that this experimental technique could be substantially improved.

4.3.4 Pendellösung Fringes

In 1968, it was suggested that a second type of crystal diffraction experiment might allow the observation time to be increased by up to a factor of 100 with little change in the effective field [101]. This technique requires large, perfect crystals in order to observe the so-called "Pendellösung" (pendulum) effect which can be described in the context of the Ewald dynamic diffraction theory.

Consider the case where neutrons are incident on a crystal at the Bragg angle in the Laue arrangement as shown in Fig. 4.6. Multiple Bragg reflection occurs and the energy or particle flow is swapped to and fro between the waves in the incident and reflected directions as the neutron travels farther into the crystal. The name Pendellösung was given to this process because of the similarity to coupled pendula. The criss-crossing waves add to give a wave traveling parallel to the Bragg planes, with $1-\cos(\pi z/\lambda \sin\theta)$ amplitude variation in the z direction normal to the Bragg planes (see Fig. 4.6). It is in fact possible to have two such eigenwaves (Borrmann waves) in the crystal, with the amplitude maximum of one corresponding to the minimum of the other. The position of the maxima and minima relative to the Bragg plane are such that one eigenwave has the maximum possible potential energy while the other has the minimum. In a monatomic substance such as silicon, with uniformly spaced planes of atoms having a positive scattering length, the wave with the higher potential energy has its maxima coincident with the atomic planes, and the wave with the lower potential energy has its maxima between the atomic planes. The difference in potential energy is on the order of the effective potential U_F (see (4.13)), i.e., of order 100 neV. Interference can take place at the exit face of the crystal giving maxima and minima in transmitted intensity, depending on the phase difference which has built up on propagation in the crystal.

A neutron EDM would be evident if there were a difference in the average contribution of $\boldsymbol{d}_n \cdot \boldsymbol{E}$ to the energies of the two eigenwaves. This would affect the interference at the exit face and would lead to a change in the transmitted neutron intensity as the neutron spin direction is changed. Figure 4.7 shows an arrangement of planes which would give a nonzero average EDM.

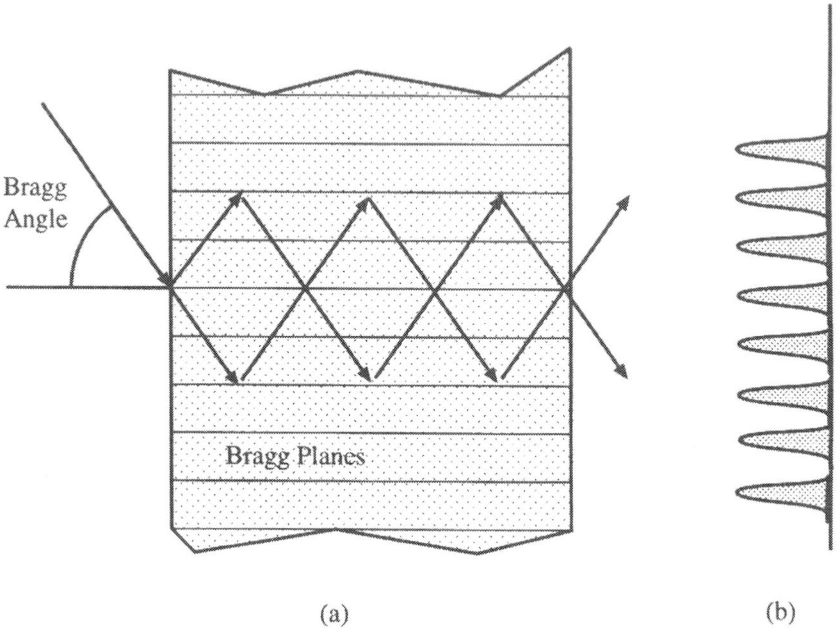

Fig. 4.6. (a) Multiple reflections for a thick crystal set up in the Laue configuration. (b) The amplitude profile for one of the eigenwaves in the crystal

The planes labeled A are of one type of atom with a large positive scattering amplitude, while those labeled B consist of another type of atom with smaller scattering amplitude. The nuclear forces are dominant in determining the positions of the amplitude maxima; for the wave of higher potential energy, the maxima will be close to the planes A shown in Fig. 4.7b. This results in a neutron density gradient across the planes B, and this gives a nonzero average EDM interaction. The averages of the two waves are opposite in sign and have magnitude of order $d_n Z_B(1-f)/10R^2$. If the nuclei have only modest absorption, a path of a few centimeters through the crystal could be used with only slight loss in transmitted neutron intensity. For neutrons with $\lambda = 1$ Å (velocity of 3×10^3 m/s), this corresponds to an observation time of 10^{-5} s.

With the availability of large quartz and other noncentrosymmetric crystals (BGO, for example), interest in this type of experiment has recently been renewed. As a prelude, the Schwinger scattering has been investigated in α-quartz by a number of workers [102, 103], limits to sensitivity have been addressed [104], and a method of controlling the motional field effect has been proposed. The expected EDM sensitivity can be estimated for diffraction from a 10 cm cube of α-quartz [$(11\bar{2}1)$ plane] using 4.4 Å neutrons; in this case, the interaction time is on the order 2 ms and the electric field is on the order of 0.2 MV/cm. With an anticipated counting rate of 400 s^{-1}, the sensitivity would be comparable to the ultracold neutron EDM experimental sensitivity (about 3×10^{-25} e cm per day of measurement). However, given that the relative field angles must be controlled at the level of 10^{-7} rad to be competitive with current neutron EDM experimental limits, there appears to be little hope of an improved systematic-free limit by using this technique. The problems which may be encountered here are similar to those expected in T violation studies in polarized neutron transmission through polarized targets, as discussed in [46]. The disadvantage as compared to other EDM experiments is that the internal field cannot be turned off as in the case of an externally applied electric field. One is then faced with the problem of aligning the spin and crystal angles to the absolute accuracy given above.

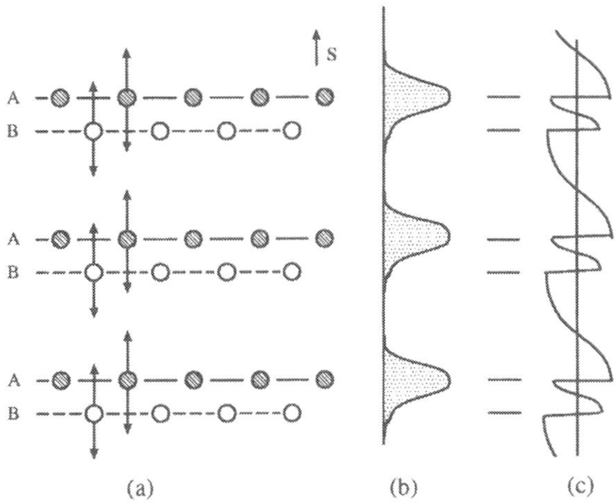

Fig. 4.7. (a) A set of Bragg planes which could be used for an EDM experiment. The nuclei of atoms A and B have large and small neutron scattering amplitudes, respectively. (b) Variation in the amplitude across an eigenwave. (c) Position variation of the EDM interaction with neutron spin perpendicular to the Bragg planes. The positive regions of the interaction are near the eigenwave amplitude maxima, while the negative regions are near the amplitude minima.

4.3.5 Neutron Beam Experiments, 1968–1973

Work on the Oak Ridge neutron EDM experiment continued, and in 1968 a new upper limit of 3×10^{-22} e cm was reported [105]. The most important result of this work was identification of the systematic effect evident in the earlier data; the source was a motion of the apparatus associated with mechanical vibrations due to the remote switching of the electric field direction.

In 1969, results of an experiment at the Brookhaven high flux beam reactor were reported [106]. This experiment appears to have been plagued with technical difficulties, perhaps due to the larger motional field effect for the higher velocity thermal neutrons (2000 m/s) as compared to the 80 m/s velocity neutrons used in the Oak Ridge experiments. A limit of 10^{-21} e cm was obtained.

Results with a new apparatus, operated at the Oak Ridge low-intensity test reactor, were reported in 1969 [107]. Improvements incorporated into this apparatus included a more stable and uniform magnetic field, better control over the angle between the electric and magnetic fields, improved magnetic shielding, a longer interaction region (2 m), and most importantly, the apparatus was routinely rotated to reverse the direction of the neutron beam through the apparatus. Insofar as the neutron trajectories and stray fields were the same, this rotation could be used to discriminate motional field systematic effects. The result was

$$d_{\mathrm{n}} = (1.54 \pm 1.12) \times 10^{-23}\ e\ \mathrm{cm}; \quad d_{\mathrm{n}} < 5 \times 10^{-23}\ e\ \mathrm{cm},$$

or nearly an order of magnitude improvement over the previous best limit. The relative electric and magnetic field angle was found to be 2.7 ± 1.6 mrad, larger than the mechanical tolerances of the pole pieces. This was presumed to be due to tangential residual components of magnetization in the pole faces resulting from imperfect demagnetization. It was anticipated that this apparatus, operated at a more intense source, could produce a limit at the level of 10^{-23} e cm.

Also reported in 1969 was the result of an experiment performed at the VVR-S water moderated reactor of the Institute of Atomic Physics, Bucharest [108]. This experiment used a well-collimated beam of thermal neutrons from the reactor horizontal channel. The result was an upper limit of 8×10^{-22} e cm for the neutron EDM, with no discernible systematics. This experiment incorporated a number of nice features, but was limited by the lack of low energy neutrons and a low neutron flux.

A beam experiment using the HERALD reactor at Aldermaston was begun in 1968, when an upper limit for the neutron EDM of about 10^{-21} e cm was obtained; work continued, and in 1973 a limit of about 5×10^{-23} e cm was obtained. These results were never published.

In 1973, additional results at Oak Ridge, using the 1969 apparatus with modifications to improve the determination of the motional field effect, were reported [109]. In addition to an improved data acquisition procedure which eliminated the effects of sparks more efficiently, and to improved polarizers based on reflection from pure iron magnetic mirrors, the entire apparatus was mounted on a surplus naval-gun mount so that it could be easily rotated, thereby reversing the sign of the motional field. The result of this experiment was

$$d_{\mathrm{n}} = (3.2 \pm 7.5) \times 10^{-24}\ e\ \mathrm{cm}; \quad d_{\mathrm{n}} < 1.0 \times 10^{-23}\ e\ \mathrm{cm},$$

and was limited by counting statistics.

4.3.6 The Institut Laue-Langevin (ILL) Experiment of 1977

The Oak Ridge experimental apparatus used for the 1973 result was moved to the ILL high flux reactor in Grenoble, France. This reactor has a liquid deuterium moderator placed close to the core, and provided a cold neutron flux approximately 1000 times that of the Oak Ridge test reactor. Again, the naval-gun turret was used to rotate the 3 m long apparatus by 180° in order to reverse the neutron velocity relative to the applied electric and magnetic fields; with the improved sensitivity, this was now a crucial experimental control. The so-called "string" method of eliminating background magnetic field drifts in the data was developed for this experiment. In fact, with the improved sensitivity, this experiment was affected to a much higher degree by sparks, leakage currents, magnetic field changes correlated with the high voltage switching mechanism, and forces between the electrodes.

After the motional field problem, the major systematic was sparking changing the magnetization of the pole faces which also served as electrodes. Such sparks were easily detected, and the contaminated data rejected.

The force between the electrodes, several Newtons with a field of 100 kV/cm, compressed the quartz spacers defining the magnet gap and distorted the magnetic pole faces slightly. The magnitude of this effect, quadratic in the applied voltage, was easily obtainable from data taken at zero electric field between each high field measurement of alternately reversed polarity. The frequency shift was about 4.5 mHz at 100 kV/cm, and was entirely attributable to the compressibility of the quartz spacers. This effect placed stringent requirements on the reversibility of the electric field, in addition to introducing concerns over preferential sparking for one particular electric field polarity over a long period of data collection.

The accuracy of the final result was not limited by counting statistics, but by instabilities in the various systematic effects. The final result, reported in 1977 [110], was

$$d_\mathrm{n} = (0.4 \pm 1.5) \times 10^{-24}\ e\,\mathrm{cm}; \quad d_\mathrm{n} < 3 \times 10^{-24}\ e\,\mathrm{cm},$$

where the uncertainty is determined largely by systematic effects. This was the final neutron beam experiment; uncontrollable systematic effects were dominating over the counting statistics, and prospects for improving the beam technique were poor.

4.4 Ultracold Neutrons

The present limits on the neutron EDM are the results of experiments employing stored ultracold neutrons (UCN); observation of total reflection of neutrons leads to the speculation that neutrons with total kinetic energy E less than U_F (the effective potential presented by a material boundary) will be reflected for all incident angles and hence can be stored in closed material vessels. Since $U_F \approx 200$ neV (see Table 4.2), UCN have velocities of order 5 m/s, wavelengths of order 500 Å, and an effective temperature of order 2 mK. Although the idea of containing neutrons is attributed to Fermi, the first person to take the idea seriously enough to put it into print was Zel'dovich in 1959 [111]. In fact, his paper begins with the statement, "The idea of retaining slow neutrons has been mentioned many times, but the corresponding experiments have not yet been performed, and the literature does not contain even rough estimates pertaining to this problem."

The relative fraction of the time that a UCN spends in contact with a wall of a storage vessel is estimated in [111] as follows. If the diameter of a hollow spherical cavity is L, the time between wall collisions is $t = L/v$, where v is the UCN velocity. Now, when a neutron undergoes reflection from a potential step, the wavefunction tunnels into the surface and remains in the material for a time of order

$$t' = \frac{\sqrt{2m}}{k_0\sqrt{U-E}} \approx 4 \times 10^{-9} \text{ s} \tag{4.39}$$

so the net neutron loss rate is the loss probability per reflection (4.10) times the number of reflections per second, giving

$$\tau^{-1} = \rho\sigma_0 v_0 t' v/L \tag{4.40}$$

which, for graphite where $\rho\sigma_0 v_0 \approx 100$ s^{-1}, gives $\tau = 2.5 \times 10^5$ s for $U - E =$ 100 neV, and $v = 5$ m/s, with L =50 cm, which is much longer than the neutron β decay lifetime (about 890 s). This is in good agreement with more detailed calculations [112, 113]; using the ratio $f = W/U_F$ given in Table 4.2 [see (4.26)], averaging the loss probability for a neutron with kinetic energy E over all incident angles gives

$$\bar{\mu}(E) = 2f\left(x^{-1}\arcsin\sqrt{x} - \sqrt{x^{-1}-1}\right), \tag{4.41}$$

where $x = E/U_F$. When $x = 1/2$, $\bar{\mu} = f$.

Although the generation, transportation, and storage of UCN are in themselves profound tests of fundamental physical principles, the development of more intense UCN sources, improved guides, etc., is currently motivated by the usefulness of UCN as a tool to determine properties of the neutron and to study its fundamental physical interactions. UCN are useful because they can be readily stored in bottles for times approaching the β decay lifetime. In a suitable magnetic environment, the spin polarization lifetime can easily exceed the storage lifetime. In addition, the low velocity (long wavelength) of UCN and the nearly ideal potentials presented by magnetic fields, gravitational fields, and material surfaces make a number of interesting experiments possible. In addition to the best limits on the neutron EDM, UCN storage experiment have also given the best value for the neutron lifetime [114].

Now let us consider the magnitudes of various interactions affecting UCN. First, UCN gain about 100 neV of energy per meter fall in the Earth's gravitational field. This energy is comparable to typical values of U_F. Next, a magnetic field of 1 T corresponds to an energy shift of ± 60 neV, where the sign depends on the direction of the neutron spin relative to the field. This has been used to produce polarized UCN; the ratio of Fe to Co in a thin foil is adjusted so that, when the foil is magnetically saturated, the net magnetic plus effective potential cancel for one spin state, whereas for the other spin state, the effective potential is doubled. Thus, UCN in the appropriate spin state will be totally reflected, while the other state passes through the foil with no attenuation, other than that due to nuclear absorption. The other force which governs the life of a UCN is the weak interaction which leads to β decay.

Early attempts at UCN storage resulted in disappointingly short storage time; the short storage effect was ultimately linked to surface contaminants, particularly hydrogen, and now storage lifetimes for carefully prepared surfaces are approaching the value of 10^5 s given above.

In [111] it was also estimated that a thermal flux of 10^{12} n/(cm^{2}s) cooled to 3 K in liquid He would produce a UCN density of 50 cm^{-3}. Such densities have now been produced at the Institut Laue-Langevin, using a reactor with thermal flux of 10^{15} n/(cm^{2}s) moderated to 20 K in a liquid deuterium cold source. However, there are some subtleties to the moderation of neutrons in liquid He, as will be addressed later, and indeed superfluid He is a much more effective moderator than one might conclude based on the estimates in [111].

The first observations of UCN were by a group working in Dubna under the direction of F.L. Shapiro [115] and independently in Munich [116], in the years 1968–69. In Munich, the primary motivation for developing UCN appears to be their application to the study of condensed matter, whereas the Dubna group was interested in UCN for fundamental experiments with the neutron [117]. It was suggested that UCN could be used to search for a neutron EDM; it was recognized that a storage experiment could give orders of magnitude higher sensitivity than a beam experiment, due to a longer interaction time, 10^2 s as opposed to 10^{-2} s. Furthermore, $\boldsymbol{v} \times \boldsymbol{E}$ systematic effects associated with nonparallel E and B fields would be greatly suppressed [118]. Since the UCN velocity is isotropically directed in the storage volume, the $\boldsymbol{v} \times \boldsymbol{E}$-generated magnetic field, which leads to systematic effects in beam experiments, is essentially eliminated. Fundamental limits to sensitivity for a variety of neutron EDM experiments are discussed in [82]; use of bottled UCN is the clear winner. It was further suggested that "if no effort were spared, a limit for the dipole length of 5×10^{-27} e cm might ultimately be reached." The present experimental limits are not too far from this level and it is expected to be reached in the next round of measurements. An entirely new technique described later in this chapter might even reach the 10^{-28} e cm level.

Two recent books discuss the field of ultracold neutron physics, and the various issues and history of the field are well described therein [112, 113].

4.4.1 Sources of Ultracold Neutrons

To determine the density of UCN which may be trapped in a cavity with wall of potential U_F, expression (4.31) can be integrated up to an energy $E = U_F$, and, for $v \ll \alpha$, the net density is

$$\rho_{\mathrm{UCN}} = \frac{2}{3}\frac{\Phi_0}{\alpha}\left(\frac{U_F}{k_B T}\right)^{3/2}, \tag{4.42}$$

which gives, for $U_F = 250$ neV (Be potential), and $T_n = 300$ K, $\rho_{\mathrm{UCN}} = 10^{-13}\Phi_0$ cm^{-3}. For the ILL reactor, with $\Phi_0 = 10^{15}$ n/(cm^{2}s), it gives $\rho_{\mathrm{UCN}} = 100$ cm^{-3}. This number can be increased by a factor of 20 to 30 by the use of a liquid hydrogen or deuterium moderator at 20 K.

Unfortunately, such densities only exist in an ideal moderator, and the extraction of the UCN always involves considerable loss. Typically, a UCN guide (polished pipe of high potential material) is used to transport UCN (as a gas) from the moderator to the experimental apparatus. In this case, the moderator serves as a source of UCN current, and the guide fills until the density at the guide entrance is equal to that at the moderator surface. There are several loss mechanisms. First, there is absorption in the moderator; in the case of a hydrogenous moderator, the mean free path of a UCN before absorption is on the order of a millimeter. Thus, the UCN emerging from the moderator must be those produced within this distance of the surface, and therefore the density at the moderator surface is much lower than that given by (4.31). If liquid deuterium is used as a moderator, the mean free path increases to about 100 cm; however, in this case, when a UCN leaves the moderator, its kinetic energy increases due to the kick of the repulsive effective potential, thereby increasing the minimum UCN energy considerably. Furthermore, this potential step results in a loss of UCN current due to the usual quantum mechanical reflection. Such reflection occurs at every "window" in the UCN extraction guide; such windows are included for safety purposes. In addition, windows absorb neutrons, and neutron guides have transmission losses.

In order to help reduce window losses, and to compensate for the potential step of the moderator, vertical extraction is used. The idea is that the extracted neutrons are guided vertically, and hence lose 110 neV per meter. Thus, the neutron velocity can be relatively high at the moderator and window potential steps, thereby reducing the $1/v$ absorption losses in the windows and reflection loss at the potential steps. In the case of the world's two strongest sources, one at the ILL and the other at the Petersburg Nuclear Physics Institute in Gatchina, Russia, the vertical heights are of order 10 m and 5 m, respectively. It is well known that deceleration by a conservative potential cannot increase the UCN phase space density (Liouville's theorem; this point is discussed in [112, 113]), so the final UCN real space density is no higher than that expected at the surface of an ideal moderator; the extraction losses, however, are substantially reduced. The extraction efficiency of these sources is of order 10%, and they produce densities of about 100 and 20 cm^{-3} respectively. These sources are discussed in greater detail in [112, 113].

4.5 Neutron EDM Measurements with Stored Ultracold Neutrons

4.5.1 Present Limits for the Neutron EDM

The two best experimental limits for the neutron EDM are based on UCN storage. One experiment, based at the Institute Laue-Langevin (ILL), has produced an upper limit of $1.2 \times 10^{-25}\, e$ cm, and the other, based at the Petersburg Nuclear Physics Institute (PNPI), an upper limit of $1.1 \times 10^{-25}\, e$ cm. Since we are more familiar with the ILL experiment, this will be described in more detail. However, the two experiments are quite similar.

Since the ILL experiment was limited by systematic effects, while the level of systematic contamination to the PNPI result is unclear, it seems imprudent to statistically combine these two experimental results. Insofar as there are no definitive criteria for determining the upper limit in the presence of systematics (the ILL limit is in fact quite conservative), which is to say that both of these limits are simply "made up", we can assume they are identical. Furthermore, given the conservative nature of the limits and their agreement, in this book we take the value

$$d_\mathrm{n} < 10^{-25}\, e \text{ cm} \tag{4.43}$$

as the experimental upper limit.

4.5.2 Stored UCN EDM Experiment at the Institut Laue-Langevin

Figure 4.8 is a schematic of the experimental apparatus described more fully in [119, 120].

The apparatus was initially used on the PN5 UCN source at the ILL and produced the result $0.3 \pm 4.8 \times 10^{-25}$ $e\,$cm [120]. In 1986, the experiment was moved to the TGV neutron turbine [121] which produces an UCN flux two orders of magnitude higher than the PN5 source; the UCN density at the turbine output is 90 cm^{-3}. The final result of this experiment is $-(3 \pm 5) \times 10^{-26}\, e$ cm [122], with statistical and systematic errors of $2 \times 10^{-26}\, e$ cm and 3×10^{-26} e cm, respectively. A description of the experiment follows.

UCN were transported from the turbine to the experiment through a stainless steel guide. At the apparatus, the UCN were polarized by transmission through a magnetically saturated 1 μm thick iron–cobalt foil. The neutron storage bottle (or cell), consisting of two beryllium electrodes 0.25 m

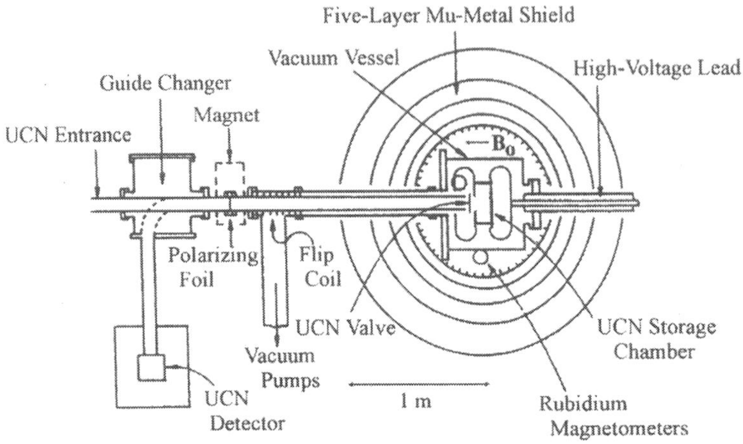

Fig. 4.8. UCN EDM apparatus used at the ILL

in diameter separated by a 0.1 m long cylindrical beryllium oxide tube (0.01 m thick wall) which served as an insulator, had a critical velocity 6.9 m/s, and volume of 5 ℓ. The edge of the cylindrical insulator rested in grooves in the electrodes (Fig. 3.3). The storage cell was inside a five-layer Permalloy shield (shielding factor of 10^5) with the bottle axis perpendicular to the cylinder axis of the shield, the orientation such that the magnetic shielding is maximum (Fig. 3.5 and Sect. 3.4.2). A 10 mG magnetic field $\boldsymbol{B}$, parallel to the axis of the bottle, was produced by a cosine distribution of currents, as shown in Fig. 3.6. Neutrons entered the bottle through a hole in the grounded electrode, which could be sealed with a beryllium door.

The magnetic field between the polarizer and the storage bottle was carefully tailored so that the adiabatic condition (4.29) was satisfied everywhere; there was a gradual change from the approximately 1 kG polarizer field to shield, and a weak field through the shield to the bottle field of 10 mG, so no significant loss of polarization occurred.

The experiment was operated as follows. The 5 ℓ storage volume was filled for 10 s (three filling time constants), after which the door was closed. Immediately after filling, the density of polarized UCN was about 10 cm^{-3}. After waiting 6 s to allow the neutron velocities to become isotropic within the bottle, the first pulse of a Ramsey-separated (in time) oscillatory field magnetic resonance sequence was applied for 4 s. This turned the neutron spins perpendicular to the magnetic field (see Sect. 3.1.2). The neutrons were allowed to precess for 70 s (the neutron storage lifetime) after which a second 4 s Ramsey pulse was applied. The neutron valve was opened and those neutrons in the appropriate spin state passed through the polarizing foil which now served as an analyzer. These neutrons were diverted by a switchable guide arrangement to a ^{3}He-doped proportional counter, and counted for 10 s. To reduce neutron losses in the detector window, the UCN were allowed to fall approximately 1 m before entering the detector. The spin-flip coil, which consisted of about 5 turns/cm wound on a NiCo-coated glass guide (so that the 3 kHz RF field could penetrate) 20 cm long, was then turned on. At some point between the shields and polarizer magnet, where the static field is varying from near zero to approximately 1 kG, the spin precession frequency was equal to the RF field frequency. The amplitude of the RF field was large enough so that the spin direction would be adiabatically reversed [see [76], Sect. IIE(c)], thereby permitting the remaining neutrons to pass through the polarizing foil and be counted, also for 10 s. The two counting periods gave approximately 12 000 and 8 000 neutrons. Including filling and emptying, each measurement cycle took 124 s. Figure 4.9 shows the sepa-

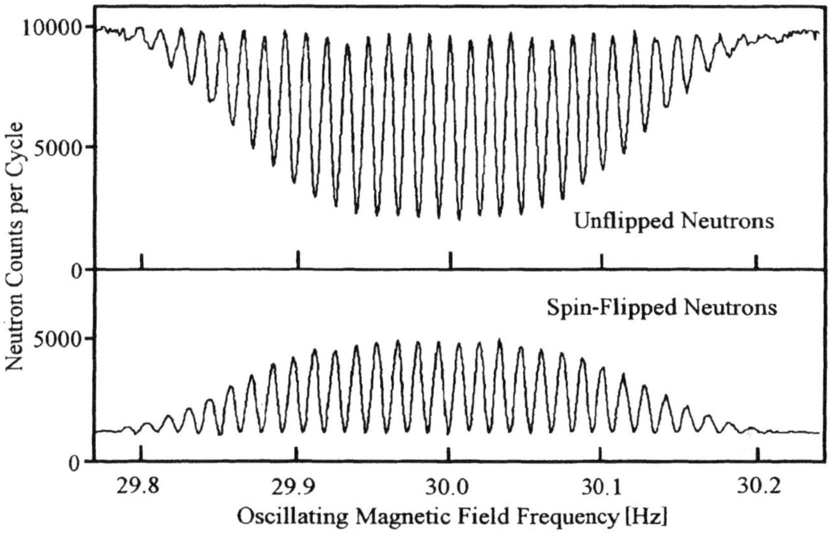

Fig. 4.9. Spin-up and spin-down neutron counts as a function of oscillating magnetic frequency with fixed static magnetic field for the ILL neutron EDM apparatus. Each point represents a measurement sequence of filling, resonance, and emptying

rated oscillatory field fringes as the frequency is varied (fixed magnetic field) between subsequent measurements. Since the average velocity is zero for all of the UCN, the fringe detail is perfect [cf. Fig. 4.3, see also (3.10)].

An electric field, $\boldsymbol{E}$, with magnitude up to 1.6 MV/m, followed a 32 measurement cycle (about one hour) sequence: The resonant frequency was determined for 8 cycles with $\boldsymbol{E}$ applied parallel to $\boldsymbol{B}$, 4 cycles off, 8 cycles antiparallel, and 4 cycles off, with 2 cycles taken to change each state (the magnitude of $\boldsymbol{E}$ was constant over each approximately 3 day set of measurements). The direction of $\boldsymbol{E}$ was reversed by simply changing the polarity of the voltage applied to the upper plate. The neutron bottle was maintained under vacuum (10^{-6} torr) or with 10^{-4} torr of nitrogen or helium to help quench sparking; it was found that helium works much better (Sect. 3.3.1). Leakage currents across the bottle were monitored and kept below 30 nA by operating at an appropriate voltage. For most of the accumulated data, the leakage current was less than 5 nA.

Over the course of a reactor cycle (about 6 weeks), the neutron storage and high voltage properties of the bottle gradually deteriorated. This was probably due, in part, to hydrogen contaminating the surface. The background vacuum was rather poor, of order 10^{-6} torr [123]. Running a discharge of Ar and D_2 *in situ* (about 150 V, 50 Hz for 10 min at 1–5 torr) restored the storage time and low leakage currents [124]. Such discharge cleaning was repeated approximately once every reactor cycle. There was some evidence that the presence of the high voltage accelerated the degradation.

The magnetic field within the shield was monitored using three optically pumped rubidium magnetometers placed around the neutron bottle at a maximum distance of 40 cm from the bottle. The field at each magnetometer was averaged over the neutron storage time. In addition, there was a flux gate magnetometer placed between the outer two layers of the shield to check for possible externally generated systematic signals.

The frequency of the oscillating field for the Ramsey pulses was changed between subsequent storage periods so that there would be a continuous calibration of the apparatus; the frequency was varied sequentially through four points near the half maximum of the central fringe of the resonance pattern. The two points on each side of the resonance center were separated by one tenth of a linewidth. The four points were used to determine both the neutron spin polarization and the resonant frequency. To determine the neutron resonant frequency, ν_n, a first pass was made through the data to extract the visibility of the resonance curve, $\alpha = (C_1 - C_2)/(C_1 + C_2)$, which was typically 0.64 (Fig. 4.9). A second pass used α and a combination of the counts for the two neutron spin states to yield a single frequency ν_n for each measurement cycle, as implied by (3.11). An important advantage of this technique was that it suppressed nonstatistical fluctuations in the neutron flux (due to reactor power fluctuations, for example). This ratio is independent of the total number of neutrons stored in the bottle. Over the course of the measurement, the frequency was changed so that the resonance was followed as the magnetic field drifted.

A run lasted about three days and had about 1000 measurements of the resonant frequency. The electric field was reversed about every 10 measurements as described above.

Figure 4.10 shows ν_n as a function of time for a typical one-day run, and the same data after the average of the magnetometer readings (normalized to the neutron frequency) was subtracted from ν_n. Jumps in the magnetic field, as shown in Fig. 4.10, might be due to, for example, movements of the reactor crane, to relaxation of the magnetic shield residual magnetization, or to the switching of magnets on neighboring experiments.

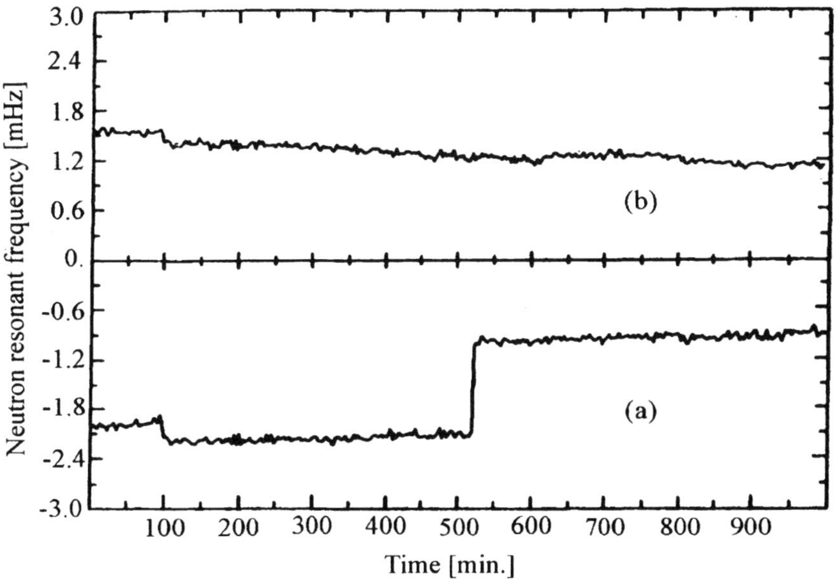

Fig. 4.10. (a) Drift in neutron resonant frequency as a function of time. (b) After subtraction of the average Rb magnetometer reading. While the large jump is substantially reduced, the small jump is only reduced by about a factor of two

An EDM would appear simply as a shift in resonance frequency with application of an electric field. The question is how to extract the EDM frequency shift from other systematic effects due to sparks, leakage currents, or externally generated fields associated with switching the high voltage.

A number of techniques were used to extract the neutron frequency shifts which change sign with the applied electric field direction, $\Delta\nu_{\rm n} = [\nu_{\rm n}(\boldsymbol{E}) - \nu_{\rm n}(-\boldsymbol{E})]/2$. Generally, a drifting background was removed by fitting out a linear and quadratic drift in resonant frequency with time for the individual measurement sets. Cycles around obvious discrete jumps in the magnetic field were discarded. Similar analyses were performed for the three magnetometers yielding $\Delta\nu_{\rm m1}$, $\Delta\nu_{\rm m2}$, and $\Delta\nu_{\rm m3}$. Frequency shifts quadratic in $\boldsymbol{E}$ (voltage on/voltage off effects) and shifts between the zero electric field groups (due for example to the shields being magnetized by leakage currents or sparks) were also extracted. To help identify systematic effects, the direction of the fixed magnetic field was reversed every few weeks; a true EDM frequency would have its sign reversed by this. In addition, the high voltage was varied between measurements sets; a true EDM would scale with the electric field strength.

For all of the data taken over the two-year running period, individual measurement set results were combined into 15 groups, one for each six week reactor cycle of measurement time. The weighted average of these data yields an uncorrected neutron EDM of $\tilde{d}_{\rm n} = (-1.9 \pm 2.2) \times 10^{-26}\, e$ cm with a normalized $\chi^2 = 3.1$ and was taken as evidence for nonstatistical (systematic) variations in the uncorrected EDMs.

The magnetometer readings were used to correct for systematic errors in two independent ways. The first made use of the observation that for perturbations external to the magnetic shields, the average value of the three magnetometers usually tracked the neutron resonance frequency to within 20% (as illustrated in Fig. 4.10). For each measurement set, a corrected $\Delta\nu_n$ was obtained by subtracting the average of $\Delta\nu_{\rm m1}$, $\Delta\nu_{\rm m2}$, and $\Delta\nu_{\rm m3}$. After correcting by the averaged magnetometer results, $d_{\rm n1} = (-3.4\pm2.6)\times10^{-26}\ e$ cm with a normalized $\chi^2 = 2.2$. The reduction in χ^2 for the corrected data came primarily from the additional noise contributed by the magnetometers. Although this technique did not eliminate the nonstatistical variations in the data, it reduced contributions to $d_{\rm n}$ from external magnetic field fluctuations.

The second magnetometer correction method allowed for the possibility that systematic frequency shifts may be different for the neutrons and magnetometers (such as those due to magnetic fields from leakage currents across the storage bottle). As above, the frequency shift with the electric field $\Delta\nu_{k,i}$ over each run (set of 1000 measurement points) where k represents the neutrons or any of the magnetometers and i represents the run number, were determined. A true neutron EDM is a frequency shift $\delta\nu_{\rm n}$ which is correlated with the electric field and direction of the magnetic field, that is $E\hat{B}$ where E is the electric field magnitude and $\hat{B}$ is the sign of the magnetic field relative to the electric field, and *which is not correlated with a shift on the magnetometers* . Any part of $\Delta\nu_{\rm n}$ correlated with the magnetometers shift is possibly a systematic effect, indistinguishable from a true EDM.

There are two cases to consider:

1. Systematic effects which do not follow $E\hat{B}$, e.g., those arising from ground loops or pressure sensitive sparks, etc. In this case $\Delta\nu_{\rm n}$ would be correlated with the magnetometers but not with $E\hat{B}$.
2. Systematic effects which are proportional to $E\hat{B}$, e.g., helical leakage currents. In this case, it is not possible to distinguish between an EDM and a systematic effect. Instead, a correlation between the magnetometers and $E\hat{B}$ indicates that a systematic shift is present. If such a correlation exists, the error on $\Delta\nu_{\rm n}$ must be increased.

For consecutive data sets between major alterations to the experimental apparatus (several reactor cycles), linear correlation coefficients between $\Delta\nu_{\rm n}$ and $\Delta\nu_{\rm m1}$, $\Delta\nu_{\rm m2}$, and $\Delta\nu_{\rm m3}$ were evaluated. A multiple linear regression routine [125] was then used to fit $\Delta\nu_{\rm n}$ versus a term linear in the magnitude of $\boldsymbol{E}$ and linear terms in any $\Delta\nu_{{\rm m}i}$ if the mi correlation with $\Delta\nu_{\rm n}$ was statistically significant (probability greater than 80%). There was essentially no systematic-free data as indicated by the cross-correlations.

Such correlations indicated the presence of systematic effects. Selection of seemingly "clean" data gave a 2σ result [126] that agreed with a $2.5\,\sigma$ result obtained in the PNPI experiment [127]. It is important to note that in order to see such correlation, it was necessary to look at groups of about 20 runs, that is, data taken over entire reactor cycles.

The weighted average of the four sets of data by this analysis yielded a second corrected value for the neutron EDM, $d_{n2} = (-3.3 \pm 3.5) \times 10^{-26}\, e$ cm with a normalized $\chi^2 = 1.5$. The larger error for d_{n2} compared with d_{n1} comes from the uncertainties generated by cross-correlations in the linear regression and from the use of a smaller data set. Most importantly, the regression made the data sets between major alterations of the apparatus consistent.

Similar analysis was done for frequency shifts "quadratic" (voltage on/ voltage off effects) in the applied field, $\boldsymbol{E}$: $\Delta\nu_Q = [\nu(|\boldsymbol{E}|) - \nu(0)]/E$ giving $|d_Q| < 2 \times 10^{-26}\, e$ cm. Again, there were highly significant correlations between the various channels which were compensated for by the regression fit. This was also true for the zero-field shifts, that is, at $\boldsymbol{E} = 0$: $\Delta\nu_0 = [\nu(\boldsymbol{E} = 0 \text{ after } +\boldsymbol{E}) - \nu(\boldsymbol{E} = 0 \text{ after } -\boldsymbol{E})]/E$ gives $|d_0| < 3 \times 10^{-26}\, e$ cm.

The data analysis rejected measurement cycles on either side of a detected spark. When the cycles with sparks were included in the analysis, no significant changes in the results are found. Effects from small, undetected sparks were therefore expected to be negligible. The size of a systematic generated by the leakage current, if that current were to flow in a loop around the BeO cylinder circumference, can also be of interest. For a current of 10 nA and a radius of 10 cm, the magnetic field is 50 pG, or a shift in frequency of 0.15 μHz. This corresponds to an EDM of roughly $3 \times 10^{-26}\, e$ cm, about a factor of 10 larger than predicted by (3.37) because the relatively large leakage current. Although this is an upper limit, one should be aware that the shift at the magnetometers could be different in magnitude and of opposite sign (relative to the neutrons) because they are outside the current loop; in fact, the systematic effect observed is well explained by the leakage currents that exist in the usual operation of the experiment.

Components of the neutron velocity, $\boldsymbol{v}_n$, that circulate around the bottle circumference could have generated a frequency shift proportional to $\langle \boldsymbol{v}_n \times \boldsymbol{E} \rangle$. Such circulation may persist for times on the order of the bottle emptying time constant (3 s). No evidence for such systematic effects was observed when the time interval (usually 6 s) between closing the bottle door and applying the first Ramsey pulse was varied. Also, as discussed in Sect. 3.5.3, the quadratic motional field effect was insignificant.

In summary, there was no evidence for systematic errors apart from the nonstatistical variation in the neutron frequency shifts. The corrected neutron EDM values, d_{n1} and d_{n2}, are in good agreement and differ little from the uncorrected value, $\tilde{d}_n$. The conclusion was that the best result for the neutron EDM comes from the average of d_{n1} and d_{n2} with the larger of the two errors (times $\sqrt{\chi^2}$):

$$d_n = (-3.3 \pm 4.3) \times 10^{-26}\, e \text{ cm}, \tag{4.44}$$

$$|d_n| < 12 \times 10^{-26}\, e \text{ cm} \quad (90\% \text{ confidence limit}). \tag{4.45}$$

The contribution to the above error from neutron counting statistics, using (3.13), is 1.9×10^{-26} e cm, corresponding to a total of 10^9 UCN.

Unfortunately, in experiments such as this one, it is impossible to make complete use of available densities of neutrons. The ILL turbine source produces, directly at the output, about 100 UCN cm^{-3}; after the storage period, the effective density in the EDM experiment was 4 cm^{-3}. This reduction in density is due in part to a finite bottle storage lifetime; at 70 s, the density is reduced by a factor of $e^{-1} \approx 1/3$. Furthermore, a direct loss of a factor of two from polarization of the UCN, and losses due to absorption in the polarizer of about 50% (which enters twice because the UCN pass through it twice) account for a factor of 2^3; the rest of the loss is due to finite transmission of the UCN guide. The efficiency of UCN use, based on these losses, was only 3%, in good agreement with the experimentally observed effective density of 4 cm^{-3}.

Nonetheless, this experiment was limited by how well the magnetic field within the neutron bottle was monitored by the spatially separated rubidium magnetometers. This problem is well illustrated in Fig. 4.10 where the large magnetic field jump was accounted for by the average magnetometer readings, while the small jump was made worse by subtraction of the magnetometer average. The implication is that the sensitivities of the magnetometers and of the neutrons depend on the source of field fluctuations, for both internal and external perturbations. In Sect. 4.5.4, a new experiment which is being built from the components of this experiment and incorporating a comagnetometer (a polarized atomic species within the neutron storage volume) will be described; this should eventually give a net factor of ten improvement in the limit on $d_{\rm n}$.

4.5.3 UCN EDM Experiment at the VVR-M Reactor, Petersburg Nuclear Physics Institute

The first result for this experiment, $d_{\rm n} < 1.6 \times 10^{-24}$ e cm, was reported in 1980 [128], and an improved result of $< 6 \times 10^{-25}$ e cm was given in 1981 [129]. These versions of the experiment operated in a continuous flow mode; the UCN storage container had entrance and exit openings which were not closed, so the effective UCN storage time was limited to about 10 s. To increase the sensitivity of the apparatus, improvements were made so that true UCN storage, such as that used in the ILL experiment, was possible. The first improved result was $d_n = -(1.4 \pm 0.6) \times 10^{-25}$ e cm, and was interpreted as an upper limit of $d_n < 2.6 \times 10^{-25}$ e cm[127]. Additional improvements led to the most recent result of $2.6 \pm 4.2 \pm 1.6 \times 10^{-26}$ e cm, the first error being the statistical limit, and the second an estimate of possible systematic contamination [130]. This is interpreted as an upper limit of 1.1×10^{-25} e cm. The major improvements in the final version were in the UCN source intensity, and an increased neutron storage time.

A schematic of the experimental apparatus is shown in Fig. 4.11. In many ways the apparatus is similar to that of the ILL experiment; however, there are important differences. As can be seen in Fig. 4.11, there are two neutron storage chambers, of 20 ℓ volume each, with oppositely directed electric fields (relative to the magnetic field) in each chamber. The high voltage is applied to the plate separating the two chambers while the outer plates are held at ground potential. The electric field was usually about 15 kV/cm, somewhat higher than the average of the ILL experiment.

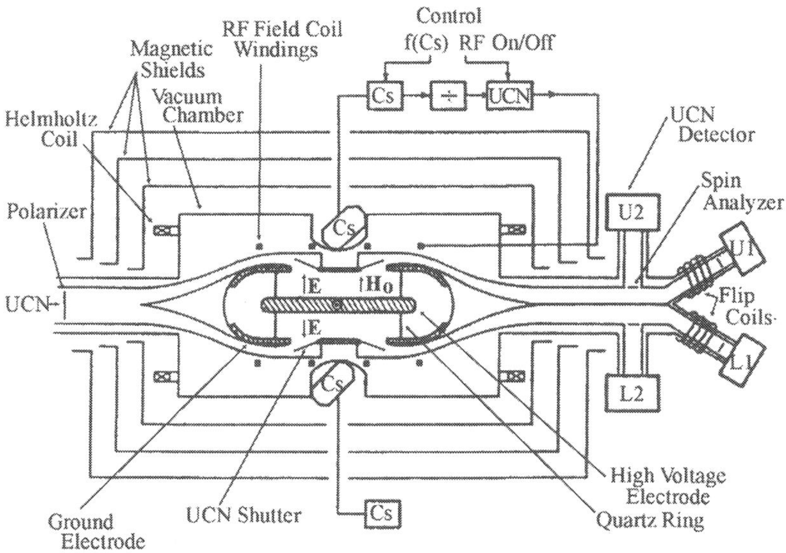

Fig. 4.11. UCN EDM experiment at the PNPI

Using two bottles with oppositely directed electric fields increases the sensitivity to a neutron EDM by $\sqrt{2}$, while reducing background magnetic field noise; an EDM generated shift will be of opposite sign for the two chambers. In addition, since the two chambers are located quite close spatially, one would expect high discrimination of background magnetic field changes; the EDM shift is given by the difference in the resonant frequency between the chambers as a function of electric field direction. This difference is sensitive only to changes in spatial gradients of the magnetic field. Such gradients could be due to locally generated fields such as leakage currents within the bottle. However, even for local fields, the close proximity of the chambers is likely to give a fairly good cancelation of net magnetic field change.

The position of the resonance was stabilized by synthesizing the radiofrequency pulses for the neutrons from the output of two cesium magnetic Zeeman atomic oscillators (located near the storage chambers). (The ratio of neutron resonance frequency to cesium is about 120.) This was necessary in part because the three-layer Permalloy shield did not provide adequate stability and shielding. The stabilization improved the effective shielding by a factor of about 15.

The chamber walls were quartz rings onto which films of BeO and Be_3N_2 had been deposited. Leakage currents averaged 30 nA at the approximately 15 kV/cm used. A system of shutters at the entrance and exit guides were used in the filling and emptying of the chambers. A system of four detectors, each with its own polarizer and two with adiabatic spin flippers allowed the spin up and spin down neutrons to be counted for each chamber simultaneously, as opposed to the ILL experiment where the spin up and spin down were counted in sequence.

As mentioned already, the experiment was originally operated in a "flow-through" mode; that is, the neutron shutters were kept open and neutrons were continuously counted; a shift between spin up and spin down counting rate as a function of electric field direction would be an indication of an EDM. It should be noted that there existed the possibility of a $\boldsymbol{v}\times\boldsymbol{E}$ shift in the flow-through system. The experiment was subsequently improved by incorporating a sequenced storage technique as was used in the ILL experiment, making full use of the chamber lifetime of 50 s. The increase in sensitivity gained here along with the improved source of UCN resulted in a daily statistical uncertainty of $\approx 2.5\times10^{-25}$ e cm, comparable to the ILL experiment.

The 1986 result now appears to have been due to an unknown systematic. In the data analysis, each bottle was analyzed separately for common-mode (overall magnetic field change) versus a differential shift. The differential change with application of the electric field led to a nonzero effect of $3.5\,\sigma$ in one chamber while it was only $1\,\sigma$ in the other. In the 1992 publication, the discrepancy between the results reported in 1986 and 1992 was attributed to a single data run in 1980. After discarding this run ($3.4\pm1.3\times10^{-25}$ e cm), the 1986 result becomes $-9\pm7\times10^{-26}$ e cm, more or less in agreement with the 1992 result.

The systematic error for the 1992 result, 1.6×10^{-26} e cm, was obtained by analysis of the Cs magnetometer data, the total effect amounting to $1.9\pm1.6\times10^{-26}$ e cm. The leakage current of 80 nA was assumed to produce a false effect of less than 0.5×10^{-26} e cm, based on the cancelation between the two storage chambers, assuming the leakage current flowed along a path 30° from the bottle axis. It is noteworthy that such high accuracy was obtained from the Cs magnetometers given that they were used primarily to stabilize the magnetic field; apparently, shifts in the gradient associated with a net magnetic field change were relatively small. In addition, another layer of magnetic shielding had been added. The reliance on the separated magnetometer signals should be contrasted with the experience gained in the ILL work where it was found that magnetic perturbations affect the UCN and magnetometers differently (see Fig. 4.10), for both external and internal perturbations. The higher statistical accuracy of the ILL experiment made it just possible to see anomalous correlations between the neutron and magnetometer signals, bringing into question the prudence of a direct subtraction. On the other hand, the quoted systematic error for the PNPI experiment is quite reasonable.

The statistical error for the PNPI experiment is twice that of the ILL experiment, while the total storage volume for the ILL experiment was 1/8 that of the PNPI experiment, and the total running times differed. The PNPI source produced a UCN density of about 15 cm^{-3}, compared to about 100 cm^{-3} for the ILL source. Also, no data taken with zero applied electric field were reported for the PNPI experiment; although such data do not contribute directly to EDM statistics, it can be quite important in the search for systematic errors, and proved to be quite useful in the ILL experiment.

A planned improvement to this experiment is the incorporate potassium magnetometers extremely close to the neutron storage chambers; the magnetometer volumes will comprise the electrode volumes and thus be in the form of thin plates [131]. Such extremely close proximity of the magnetometers to the UCN storage volume should give additional discrimination against systematic magnetic fields.

4.5.4 The ^{199}Hg Comagnetometer UCN Experiment

Since the ILL experiment described in Sect. 4.5.2 was no longer limited by counting statistics but by systematics, it was decided to rebuild the apparatus and include a comagnetometer – that is, a polarized atomic species within the same storage volume as the neutrons – and thus provide a nearly exact spatial and temporal average of the magnetic field affecting the neutrons over the storage period. The use of polarized ^{3}He had already been considered [132], but the extreme difficulty in the detection of the ^{3}He polarization makes its use impractical.

The use of ^{199}Hg was suggested in 1986 [63], and more recently described in [133]. The advantage is that ^{199}Hg can be readily directly optically pumped and its polarization optically detected with 254 nm resonance radiation, as described in Sect. 3.2. Since ^{199}Hg is a $^1S_0$ atom, its ground state polarization is specified by the nuclear angular momentum, which is 1/2 for ^{199}Hg. In addition, the room temperature vapor pressure of Hg is more than adequate to provide the necessary density.

Of course, to be useful for a comagnetometer, it must be demonstrated that the chosen atomic species does not have an EDM of its own which could possibly mimic or mask a neutron EDM; in the case of ^{199}Hg, experimental limits have already been set at the level of sensitivity needed [80]. In these experiments, ground state spin-polarization lifetimes in excess of 100 s. were routinely achieved in cells of about 5 cm^3 volume, even in the presence of electric fields up to 15 kV/cm. However, these cells included 250 torr of nitrogen to improve the high voltage stability.

An unfortunate disadvantage of ^{199}Hg is that the walls of the container must be specially prepared to have long spin relaxation times. In all previous experiments, hydrocarbon waxes were used; these of course would be unusable with UCN. In addition, the wall coating has to be stable under the application of high voltage in vacuum since a high-pressure background gas cannot be used with the UCN.

Choices for a suitable wall coating for this experiment are rather limited. The fluorinated liquid "Fomblin" used in the neutron β decay lifetime as a wall coating in fact has fairly good polarized ^{199}Hg storage properties, with a relaxation time of order 15 s/cm mean free path. Unfortunately, it is quite unstable under high voltage application in vacuum. It has been shown that for a material to have high vacuum voltage stability, its vapor pressure must increase slowly with temperature, as discussed in Sect. 3.3. The vapor pressure of Fomblin certainly rises rapidly with temperature, hence its use as a diffusion pump fluid. Also, a liquid will flow to regions of electric field maxima, and such movement could lead to systematic or other deleterious effects. Another possible wall coating material, deuterated polystyrene (DPS), a plastic, was developed quite early in this experiment, but also proved to be unsuitable. The problem here is that under application of high voltage, hydrogen (deuterium) would be liberated from the material due to both large and small electrical discharges. Atomic hydrogen adsorbed on the storage walls is very effective at relaxing the polarization of the ^{199}Hg, as was demonstrated in small storage cell experiments. In the end, Teflon appears to be the most suitable wall coating material; the fused silica cylinder is left uncoated, while all metal parts exposed to the Hg vapor are coated with a thin layer of Teflon, applied by melting a layer of colloidal material onto the surfaces. The Hg polarization cell is coated with Fomblin. There is still some trouble with the high voltage affecting the Hg polarization lifetime; so far, it appears that adding oxygen along with the helium background gas (to improve the high voltage stability) helps stabilize the polarization lifetime.

A schematic of the experimental apparatus is shown in Fig. 4.12. To increase the experimental sensitivity through storage time and UCN number increases (to hopefully at least compensate for the loss of UCN density due to the lower effective potential of a materials suitable for the storage of polarized ^{199}Hg as compared to Be–BeO), a larger (about 10 times) volume storage bottle has been constructed. The magnetic shields are the same as those used in the previous ILL experiment, only the innermost layer has been removed. The loss in shielding factor is expected to be made up for by the improved volume magnetometry. In the previous experiment, there was no indication of a magnetic field gradient induced UCN polarization loss rate; however, with the larger storage vessel and diminished shielding, some gradient effects are now evident for the relatively slowly moving UCN [see (3.28)].

Since there is a considerable shift in the center of mass between the UCN gas and atomic gas in the gravitational field (due to the difference in tempera-

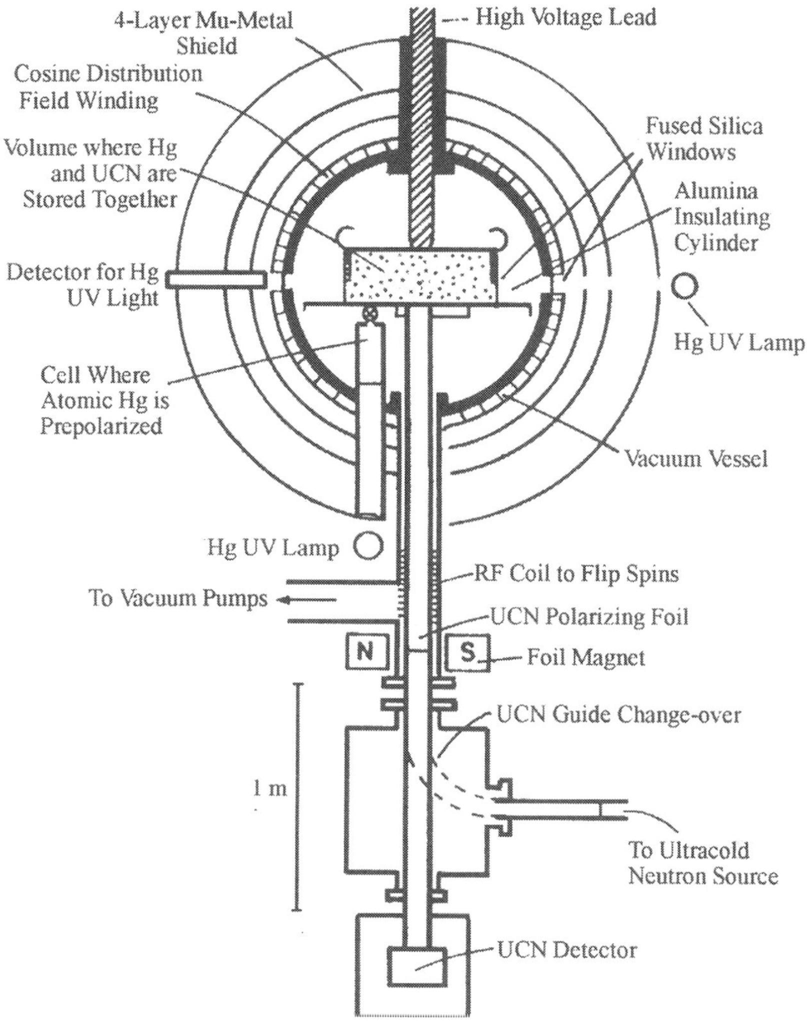

Fig. 4.12. Schematic of the ILL UCN EDM experiment incorporating a ^{199}Hg comagnetometer

ture) [132], the experiment is designed so that the shorter axis of the bottle is vertical, thus minimizing the displacement. Although the UCN gas does not strictly represent an equilibrium system, we can estimate the displacement by assuming a UCN temperature of 2.5 mK;

$$\Delta h = -\frac{h}{2} + \frac{1}{h}\int_0^h x\mathrm{e}^{-m_{\mathrm{n}}gx/kT}\mathrm{d}x \approx -\frac{m_{\mathrm{n}}gh^2}{3kT},$$

where h is the cell height. For a 20 cm cell, the displacement is on the order of 6 mm.

In addition, there is a safety consideration for the use of Hg, and it is necessary to isolate the experiment with a gas-tight window which can withstand atmospheric pressure. It was decided that this would be the polarizer; the design for the polarizer is iron evaporated onto an aluminum foil. To account for the fairly high effective potential of the aluminum, after passing through the foil, the UCN will rise about one meter. Tests were performed to determine the optimum height to maximize the number of UCN left in a test bottle after a 100 s storage period.

The neutron storage bottle is a fused silica cylindrical spacer about 60 cm diameter and 20 cm high separating aluminum plates coated with Teflon. The neutron valve is gas-tight to minimize the loss of the polarized ^{199}Hg.

Provisions are included for polarizing the atomic vapor; an optical pumping cell is connected to an isotopically enriched Hg reservoir, a few mg of HgO powder in a tube held at about 250 °C. This provides a current of ^{199}Hg atoms and can be controlled with a valve. The Hg is optically pumped to the appropriate spin state, parallel to the static field, with circularly polarized light from a Hg discharge lamp. After the Hg is polarized, and after the storage vessel is filled with polarized UCN, the neutron valve is closed; then polarized Hg is admitted to the neutron bottle. $\pi/2$ pulses are applied first for the ^{199}Hg and then for the UCN (the ^{199}Hg magnetic moment is about one third of the neutron magnetic moment). The free precession of the Hg spin is observed with a beam of circularly polarized resonance light which propagates across the bottle diameter, through fused silica optical windows in the insulating cylinder. The storage vessel spatial average magnetic field, averaged over the measurement time, can be determined from the free precession signal.

At the end of the storage period, the second neutron pulse is applied, the bottle door opened, and the neutrons are counted as usual. The Hg is pumped away. While the storage was in progress, more Hg had been admitted to the optical pumping cell and polarized; the process is thus ready to be repeated.

An EDM would be evident from a change in the ratio of the magnetic moments between reversals of the electric field. Although the sensitivity of the Hg to a magnetic field is only 1/3 that of the neutron, the high signal to noise inherent in the free precession signal is a compensating factor, and currently the determination of the average field is a factor of 3 to 10 higher in sensitivity than the neutron signal and hence contributes very little noise. The final uncertainty for this experiment, based on the shot-noise estimate given by (3.13) is about 7×10^{-27} e cm, or nearly a factor of ten smaller than the present systematic-error-dominated limit.

It is expected that this experiment will be relatively immune from systematic effects. The shift in center of mass for the UCN vs. Hg, due to the differences in their effective temperatures, is expected to give a very small percentage error in average field measurement. The magnetic fields associated with the leakage currents need only be corrected for at the 10% level of accuracy to achieve the full sensitivity. The $\boldsymbol{v} \times \boldsymbol{E}$ problem, discussed in Sect. 3.5.3, is not expected to be troublesome at the anticipated level of accuracy. It is also expected that the Hg will stick to the walls for a finite period of time because the surface presents an attractive chemical potential, contributing an extra surface average compared to the UCN. The time is estimated to be less than 1 μs at each wall collision for binding energies of order 0.1 eV [134], or a fractional time of 1/1000. We should not forget the quantum mechanical delay time for a UCN on reflection from a potential barrier; this time is of order 4×10^{-9} s [see (4.39)], so is quite insignificant.

Since the ^{199}Hg coherent scattering length has a spin dependence, the Hg gas will present an effective magnetic field to the UCN (the so-called "pseudomagnetic field") through the spin dependence of U_F. The primary effect will be due to a Hg polarization along the applied magnetic field if the $\pi/2$ pulse is not perfect. The full effective field is, assuming a Hg pressure of 10^{-5} torr, of order 400 nG, considerably larger than the level of required field control (0.05 nG). If the $\pi/2$ pulse is accurate to 1%, and the accuracy of subsequent filling pressures is within 10%, the noise contributed by this spurious field is of order 10^{-25} e cm, below the cycle counting statistics noise. One concern is that there could be a polarity-dependent Hg surface sticking time, due to, for example, more sparking in one polarity than another. This could give an electric field polarity dependence to the Hg volume average, thereby generating a systematic effect. However, at the anticipated level of accuracy, such an effect appears to be too small to be of consequence. Also of concern is a modification to the diamagnetic correction of the Hg atom during its dwell on the wall, the idea being that the electron density at the nucleus is affected by the interaction with the storage container walls. Since the diamagnetism results from the inner electrons, this effect is expected to be quite small.

4.6 The Future: Superfluid He Neutron EDM with a ^{3}He Comagnetometer

4.6.1 The Production of UCN in Superfluid ^{4}He

To increase the sensitivity of UCN EDM experimental searches, an increase in UCN density is required. Both planned and existing UCN sources, based on extraction of UCN from a cold moderator, are limited by the phase space density of low energy neutrons in the moderator. At most, one could expect a factor of perhaps ten over the density at the ILL reactor, but this requires the extension of reactor technology by about an order of magnitude in regard to radiation fluxes in the core. Spallation sources will probably not exceed the densities of currently operating reactors.

There is another way to produce UCN; the idea is to inelastically scatter cold neutrons in a suitable material. As was mentioned earlier, as the neutron wavelength increases, the inelastic scattering efficiency of solids and liquids decreases. Thus, the rate of scattering from high to low energy can exceed the inverse, that is, scattering from low to high energy. In a suitable material, the density of low energy neutrons can be enhanced over what is expected from the source phase space density.

An ideal material for such a UCN source is superfluid ^{4}He [135]. ^{4}He has zero neutron absorption because it is the most tightly bound nucleus. Thus, if the superfluid bath is sufficiently free of ^{3}He (which has a rather large absorption cross section), UCN can be stored in the bath until β decay, wall absorption, or upscattering occur; it is expected that with modest effort, β decay can become the dominant loss mechanism.

The production of UCN by the downscattering of 8.9 Å neutrons in superfluid He has now been demonstrated and well studied [136, 137]. The process is described in detail in [112]. Figure 4.13 shows the free neutron dispersion

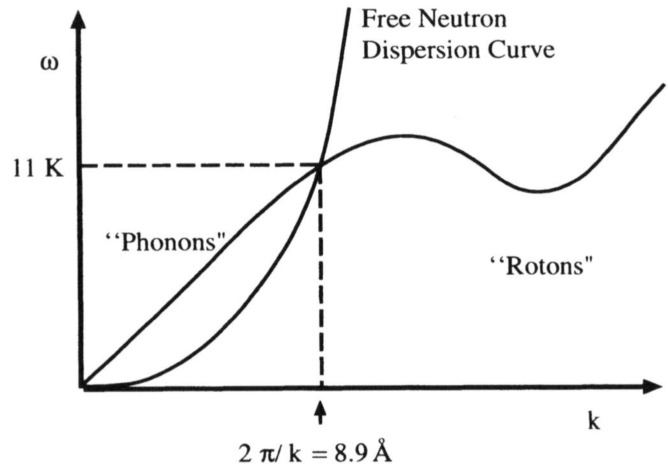

Fig. 4.13. Free neutron and superfluid ^{4}He elementary dispersion curves

curve along with the dispersion curve for elementary excitations in superfluid ^{4}He [the Landau–Feynman (L-F) dispersion curve]. The dispersion relation of the free neutron, relating the energy to the momentum, is a parabola:

$$\omega = \hbar k^2/2m. \tag{4.46}$$

This curve crosses the L-F dispersion curve at $2\pi/k^* = 8.9$ Å, and $E^* = \hbar\omega = 11$ K. The crossing point is in the linear region of the L-F curve. (The curves also intersect at $k = 0$). Neutrons in this range of wavelength are readily produced by a liquid deuterium or liquid hydrogen moderator.

Since both energy and momentum are conserved in the scattering process, neutrons at or near rest can only absorb phonons of energy E^*, where the dispersion curves cross. This process is strongly suppressed by the Boltzmann factor, $e^{-E^*/T}$, when the superfluid temperature is less than 1 K. By the same argument, only neutrons with energy near E^* can scatter into the UCN energy region by emission of a single excitation.

A UCN source based on this process operates by the following principle. Neutrons of wavelength 8.9 Å can easily penetrate the walls of the storage container, and enter the superfluid bath. These neutrons then downscatter, producing UCN which are trapped in the container. (U_F for liquid ^{4}He is about 20 neV, much less than the potential for most solid materials, so we can assume it is zero in the following discussion.) UCN produced in this way will remain in the superfluid He bath until they are lost through one of the possible loss mechanisms, which include β decay, absorption by ^{3}He, and loss in the wall. The UCN will reach a saturation density

$$\rho_{\mathrm{UCN}} = P\tau, \tag{4.47}$$

where τ is the total loss rate,

$$\tau^{-1} = \tau_{\mathrm{wall}}^{-1} + \tau_{\beta}^{-1} + \tau_{^3\mathrm{He}}^{-1} + \dots \tag{4.48}$$

and P is the UCN production rate [UCN/(cm^3 s)] due to the above-mentioned downscattering process [138]:

$$P = 7.2 \frac{\mathrm{d}^2\Phi^*}{\mathrm{d}\lambda\, \mathrm{d}\Omega} \frac{1}{\lambda_{\mathrm{u}}^3} \delta\Omega \ \mathrm{UCN\ cm^{-3}\ s^{-1}}, \tag{4.49}$$

where the neutron spectral density is specified at 8.9 Å, λ_{u} is the shortest UCN wavelength that can be stored in the container, and $\delta\Omega$ is the source solid angle subtended at the superfluid bath. A UCN source based on this principle is referred to as a "superthermal source".

The neutron–superfluid ^{4}He system is in some sense a two-level quantum system, and the production of UCN by the emission of a phonon can be compared to the spontaneous emission of radiation by an excited atom.

Cold neutrons of wavelength 8.9 Å have an attenuation length of order 100 m in superfluid ^{4}He at temperatures at around 1 K. Thus, for any conceivable experiment, the production rate will be constant, independent of position along the incident neutron beam.

The increase in neutron density near zero energy can be understood by the following argument. If we take a linear dispersion relation for the liquid He elementary excitations, $\omega = ck$ where c is the phonon (sound) velocity, we have the following condition, by conservation of energy and momentum, limiting the region around $k = k^* + \delta k$ which can scatter to an UCN with momentum k_{UCN}:

$$c|\boldsymbol{k} + \boldsymbol{k}_{\mathrm{UCN}}| = \frac{\hbar}{2m}(k^2 - k_{\mathrm{UCN}}^2). \tag{4.50}$$

The maximum and minimum of $|\boldsymbol{k} + \boldsymbol{k}_{\mathrm{UCN}}|$ are $k \pm k_{\mathrm{UCN}}$. We thus arrive at

$$\delta k = 2k_{\mathrm{UCN}}. \tag{4.51}$$

This is a remarkable result, and shows that Liouville's theorem, which was previously briefly mentioned, is apparently violated by this system. Incident neutrons occupy a (momentum) phase space volume of $4\pi k^{*2}\delta k$ whereas the UCN occupy a volume $\frac{4\pi}{3}\ k_{\mathrm{UCN}}^3$, which represents a factor of $\frac{1}{3}(k_{\mathrm{UCN}}/k^*)^2$ decrease in phase space volume, corresponding to an increase in phase space density. Given an arbitrarily long storage lifetime of the UCN, for any nonzero production rate P, the real space density will simply continue to increase as the incident neutrons downscatter, at least until the UCN density is so high that all the states of the Fermi gas are occupied, at which point no more downscattering can occur. This is possible because the produced phonons occupy a very large phase space, and these phonons are continually removed from the system by a refrigerator which keeps the superfluid bath cold.

We have not addressed upscattering of UCN by phonons which leads to additional losses. The one phonon process is easy to calculate. By using microscopic reversibility [92] the production and upscattering processes can be related:

$$\frac{\sigma(E^* \to E_{\mathrm{UCN}})}{E_{\mathrm{UCN}} e^{-E_{\mathrm{UCN}}/T}} = \frac{\sigma(E_{\mathrm{UCN}} \to E^*)}{E^* e^{-E^*/T}}, \tag{4.52}$$

which implies that the reverse process is exponentially small, as was previously mentioned. However, this simple treatment does not include higher order processes, and in fact the dominant process below 1K is two-phonon upscattering [139], which gives a loss time of about $\tau = 0.01T^7$ s.

Given that a UCN source based on this process requires a cryogenic apparatus, close placement to an intense neutron source raises the issue of radiation heating due to fast neutrons and gamma radiation. At present, it appears that a spallation source will be considerably better than a reactor in this regard, and in addition has somewhat more relaxed safety requirements. A schematic of a possible geometry for a spallation UCN source is shown in Fig. 4.14 [140, 141]. As shown, a conical shield of liquid-nitrogen-cooled water-quenched bismuth shot [142] can be used to shield the cryogenic apparatus from fast neutrons and gamma radiation; the Bragg cutoff for bismuth is about 6 Å, and if it is cooled to 77 K, the attenuation length for 8.9 Å neutrons is about 1 m. It is necessary to use highly polycrystalline bismuth in this application to ensure strong Bragg scattering at all angles, hence the use of the water-quenched shot.

Taking the expected flux from a planned 5 MW spallation source, with a solid angle of 0.03 sterad, a production rate of about $200/(\mathrm{cm}^3\ \mathrm{s})$ should be possible, and if the UCN losses are dominated by β decay, the ultimate UCN density will be about $2 \times 10^5/\mathrm{cm}^3$, or 2000 times that of existing UCN sources. This figure should be compared to a production rate of $15/(\mathrm{cm}^3\ \mathrm{s})$ in a superthermal source placed at the end of a guide.

Also of interest is a proposal to use a subcritical assembly of ^{235}U driven critical by a pulse from a small accelerator, the system being tailored to UCN production by the superthermal process [143].

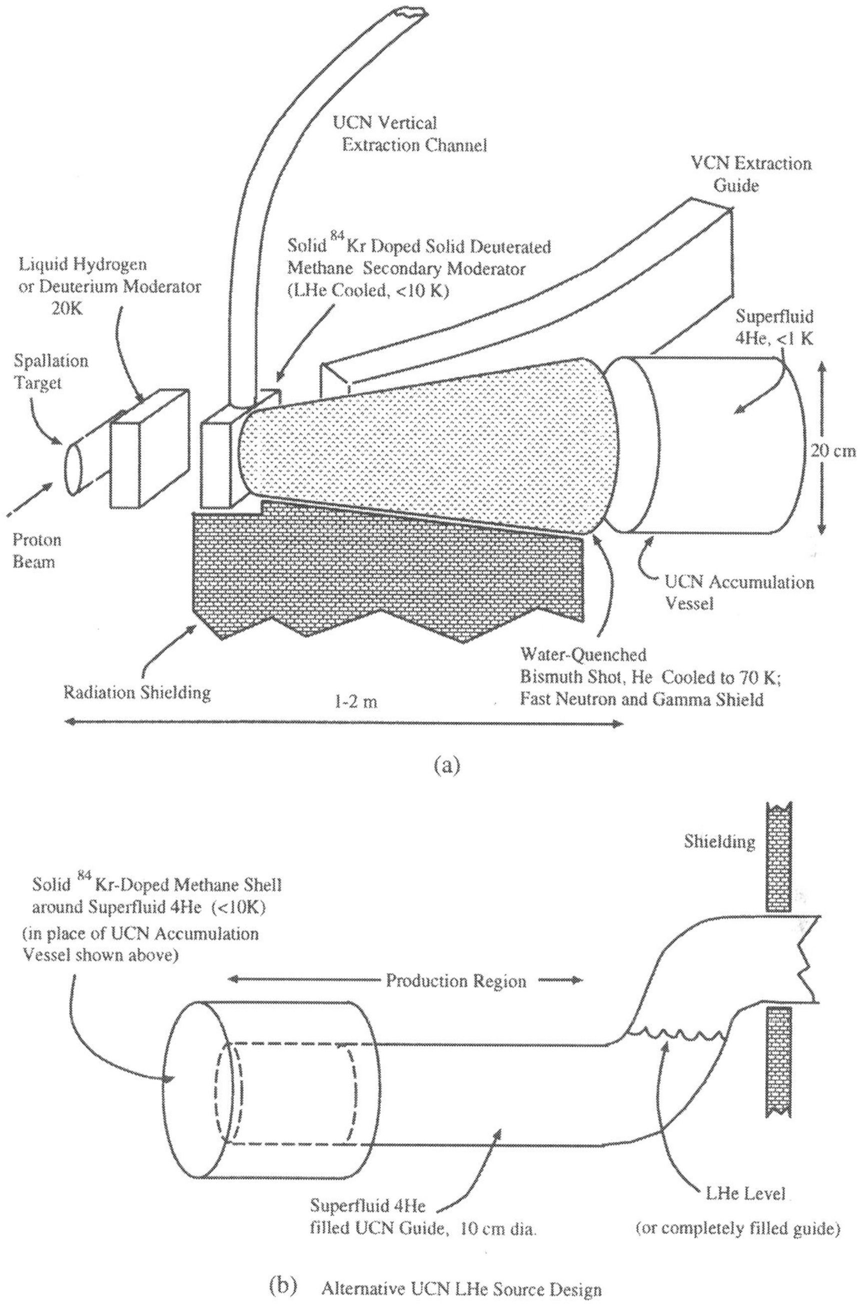

Fig. 4.14. (a) A possible geometry to generate an extremely high UCN density in a superfluid ^{4}He bath. (b) An alternative arrangement which could lead to about a factor of three increase in density; slightly faster neutrons penetrate the Bi shielding more effectively, then scatter from the moderator sheath into the superfluid bath

The UCN density produced in a superthermal source could be increased by using a "supercold moderator". For liquid hydrogen or liquid deuterium sources, the flux is maximum near 6 Å, and drops rapidly in the region of 9 Å, as was discussed in Sect. 4.2.2, because the long wavelength neutrons cannot effectively excite vibrations in liquid. However, in the case of solid methane, it is possible to reactivate the rotational degrees of freedom by doping the solid with about 20% krypton. With such a secondary supercold moderator, an increase in the density of UCN by a factor of three or more might be possible [141].

However, the most effective method of UCN production might be to use the superthermal process directly in the EDM storage cell, with the 8.9 Å neutrons delivered to the cell by a supermirror polarizing guide. In this method, the effective solid angle is about 0.01 sterad, compared with 0.03 sterad for production vessel close to the neutron source. With the guide feeding the experimental volume directly, losses associated with the Bi radiation shielding and with UCN transport in a guide, and complications with a cryogenic UCN valves, are eliminated. Furthermore, the EDM cell could be surrounded by cavity made from a high albedo material such as BeO (low loss, high incoherent cross section); this would trap the 8.9 Å and perhaps give a factor of up to four increase in the UCN production rate.

4.6.2 Superfluid ^{4}He Neutron EDM Search with a ^{3}He Comagnetometer

Attempts to make a UCN source based on the superthermal process in ^{4}He have so far encountered technical difficulties, primarily in regard to extraction of the UCN from the bath. Invariably and inevitably, the thin material windows used to contain the liquid He but allow the UCN to pass, become covered by condensed, frozen gases (O_2 or N_2), increasing U_F and/or the UCN absorption. Typically, extracted densities have been a factor of 10 to 100 below that expected [137, 144]. Indirect measurement through the upscattering rate has confirmed that the expected density does indeed exist within the bath [145].

The extraction problems can be avoided by performing an EDM search directly in the liquid helium of the superthermal source. Such a system has a number of advantages; for example, because of the excellent dielectric properties of liquid helium, increasing the applied electric field by nearly an order of magnitude might be possible. Also, it appears feasible to use a dilute solution of polarized ^{3}He as a UCN polarizer, spin analyzer, detector, and magnetometer. ^{3}He absorbs neutrons mostly when the total spin is zero because the reaction is dominated by the the 0^+ excited state [146], as follows

$$^3\mathrm{He} + \mathrm{n} \to \mathrm{p} + \mathrm{T} + 764\ \ \mathrm{keV}. \tag{4.53}$$

The polarization and cryogenic transport of polarized ^{3}He have also been studied [147]. Furthermore, energetic charged particles produce ultraviolet scintillations in liquid helium with about 4 photons per keV of deposited energy. The reaction between ^{3}He and neutrons in the liquid helium are thus easily detected, giving a detection of reactions with nearly 100% efficiency. See [148] for an application of these techniques to a measurement of the neutron β decay lifetime which may lead to a factor of 100 improvement in accuracy.

The ^{3}He serves as a UCN polarizer by absorbing neutrons in the singlet state. To be effective, this rate of absorption should be slightly higher than other UCN loss mechanisms in the system. This implies a ^{3}He concentration of 10^{-9}, or about 10^{13} atoms/cm^3. Such low densities of polarized ^{3}He can be produced with a hexapole state selector, with essentially perfect polarization. Other techniques for producing higher densities have polarization limited to 70%; since the ^{3}He serves three functions, it is expected that the experimental sensitivity varies as at least the cube of the ^{3}He polarization. This agrees with detailed calculations.

An EDM experiment based on these ideas, as originally proposed, could be sensitive to a neutron EDM by looking at the scintillation rate at the end of a double-pulse sequence, as a function of electric field polarity [149]. It has been shown by solving the Schrödinger equation with a spin dependent absorption probability that this technique is slightly less sensitive that the conventional bottle technique, however, this loss of sensitivity is more than made up for by elimination of the extraction losses and increase in electric field.

In the following discussion, let the subscript 3 refer to the ^{3}He atoms, and subscript n refer to the UCN. In the case where both species are polarized, the spin-dependent loss rate can be written

$$\frac{1}{\tau_{\mathrm{abs}}} = \frac{1}{\tau(^3\mathrm{He})}(1 - \boldsymbol{p}_{\mathrm{n}} \cdot \boldsymbol{p}_3) = \frac{1}{\tau(^3\mathrm{He})}(1 - p_{\mathrm{n}} p_3 \cos\theta_{\mathrm{n}3}), \tag{4.54}$$

where $\theta_{\mathrm{n}3}$ is the angle between the spin polarization vectors and $|\boldsymbol{p}_{\mathrm{n},3}| \leq 1$. Each loss (nuclear reaction) produces a scintillation pulse; the scintillation rate thus becomes a measure of the angle between the polarization vectors.

One could search for a neutron EDM by using the above UCN production/polarization technique. After the UCN are polarized (along a static field of magnitude B_0), the UCN and ^{3}He spins could be flipped by $\pi/2$; the spins then precess about the static field and there will be a modulation in the scintillation rate:

$$\phi(t) \propto (1 - \boldsymbol{p}_3 \cdot \boldsymbol{p}_{\mathrm{n}}) = 1 - p_3 p_{\mathrm{n}} \cos[(\gamma_3 - \gamma_{\mathrm{n}}) B_0 t + \Phi], \tag{4.55}$$

where $\phi(t)$ is the time-dependent scintillation rate, Φ is an arbitrary phase, and the gyromagnetic ratios are

$$\gamma_{\mathrm{n}}/2\pi \approx -3 \text{ Hz/mG}$$

and

$$\gamma_3/2\pi \approx -3.33 \text{ Hz/mG}.$$

The EDM of ^{3}He is expected to be quite small (due to shielding as described in Sect. 6.1); thus, if an electric field is applied along B_0 there will be a change in the frequency of the scintillation rate modulation. Unfortunately, the problem of measuring the magnetic field remains (although the effects are only 1/10 as large since the gyromagnetic ratios are nearly equal) and it has been demonstrated that experiments are presently limited by magnetic systematic effects. It might be possible to use SQUID magnetometers to detect the precessing ^{3}He magnetization, so that the ^{3}He could then serve as a direct magnetometer. Recent advances in SQUID technology make this a possible alternative to the dressed spin technique described below.

4.6.3 Dressed Spin Magnetometry

In the above description, it is evident that a perfect experiment would be possible if the magnetic moments of the ^{3}He and neutron were equal; the fact that the magnetic moments are equal to within 10% reduces the sensitivity to background field by an order of magnitude, and if the moments were exactly equal, there would be no effect at all. Unfortunately, we have no direct control over the physics responsible for the observed magnetic moments; however, these moments can be artificially modified by using "dressed atom" techniques [47], and it is possible make them equal [112].

In the presence of a strong oscillating magnetic field, the magnetic moment will be modified, or "dressed", yielding an effective gyromagnetic ratio

$$\gamma' = \gamma J_0(\gamma B_{\mathrm{RF}}/\omega_{\mathrm{RF}}) = \gamma J_0(x), \tag{4.56}$$

where γ is the unperturbed gyromagnetic ratio, B_{RF} and ω_{RF} are the amplitude and frequency of an applied oscillating magnetic RF field, and J_0 is the zeroth-order Bessel function. This effect can be qualitatively understood by taking the average of the spin in an oscillating magnetic field. Consider a spin pointing along $\hat{z}$ at $t = 0$. Now apply an oscillating field along $\hat{x}$; the spin precession frequency is time dependent,

$$\omega(t) = \dot{\theta(t)} = \gamma B_x(t) = \gamma B_{\mathrm{RF}} \sin \omega_{\mathrm{RF}} t,$$

so that the angle relative to $\hat{x}$ is

$$\theta = \gamma(B_{\mathrm{RF}}/\omega_{\mathrm{RF}}) \cos \omega_{\mathrm{RF}} t.$$

The average spin projection $\langle P_z \rangle$ along $\hat{z}$ is given by

$$\langle P_z \rangle = \frac{1}{T} \int_0^T \cos(\gamma B_{\mathrm{RF}} t \cos \omega_{\mathrm{RF}} t) \mathrm{d}t = J_0(\gamma B_{\mathrm{RF}}/\omega_{\mathrm{RF}}) = J_0(x).$$

A more sophisticated treatment shows that a spin will respond to a small (compared to the oscillating field amplitude) static field along $\hat{x}$, with an average magnetic moment $\gamma' = \gamma J_0(x)$; our simple estimate gives a picture of how the oscillating field dilutes the magnetic moment.

In practice, the oscillating field is at right angles to the static field B_0 around which the spins are precessing. In the absence of the oscillating field, one would see scintillation due to reactions occurring at a rate given by (4.55). Thus, there is an oscillation in the scintillation rate at the difference in the precession frequencies $[\delta\omega = (\gamma_n - \gamma_3)B_0]$. If the RF dressing field is now applied, the effective magnetic moments become modified, and

$$\delta\omega = [\gamma_n J_0(\gamma_n x) - \gamma_3 J_0(\gamma_3 x)]B_0. \tag{4.57}$$

This has the property that $\delta\omega = 0$ when $\gamma_n x \approx 1.19$; this condition is referred to as "critical dressing". It can be achieved in practice with a dressing field frequency of order 1 kHz and amplitude of 100 mG, and a spin precession frequency on the order of a few Hz.

If the neutron EDM is nonzero, the neutron precession frequency will be shifted by an amount $2d_n E J_0(\gamma_n x)$ (since the dressing dilutes the net spin projection). Thus, the value of $x = x_c$ to give $\delta\omega = 0$ is changed. By measuring the value of x_c vs. electric field direction, a neutron EDM would be evident. The important point is that the effect of static magnetic fields is cancelled.

Experimentally, the neutron and helium spin vectors could be kept nearly parallel; the scintillation would increase or decrease as x is varied away from the value x_c such that $\delta\omega = 0$. Over the course of a storage period, x could be sinusoidally modulated at a low frequency ω_m and the value $x_c(\pm E)$ inferred from variations in the scintillation rate which occur at harmonics of ω_m. If the average value of $x \neq x_c$, there will be a first harmonic to the scintillation rate growing linearly in time. If $x = x_c$, there will be only a second harmonic component. In practice, a feedback system might be used to force the first harmonic signal to zero; the second harmonic then serves as a system calibration. (Note that the modulation in x and the subsequent modulation in the scintillation rate are 90° out of phase because the spin vectors must precess before the effects due to a change in x are manifest).

A detailed analysis of this system is given in [81], in which many technical issues are addressed. It is shown that a factor of over 1000 improvement in the neutron EDM experimental limit is feasible. This improvement is based on a factor of five increase in electric field strength, an increase in the net UCN storage lifetime, and an increase by a factor 10^4 in UCN density. Magnetic field noise and systematic effects are eliminated by the dressed spin technique.

An important difference between the previous UCN storage experiments and one performed directly in the superfluid bath is that essentially all of the UCN stored in the liquid helium contribute to the measurement. As was shown earlier, the ILL experiment UCN-use efficiency was only 3%, giving an effective experimental density of 4 cm^{-3}. This number should be compared to the 2×10^5 cm^{-3} given above for the superthermal source.

4.6.4 Analysis of the Dressed Spin System and Systematic Effects

The motion of a spin under the application of static and nonresonant oscillating magnetic fields is quite complicated. In some sense, saying that the magnetic moment is modified (or dressed) is the "zeroth-order" approximation. In [81], the system was studied both by numerically integrating the equations of motion, and through quantum perturbation theory. The ^{3}He–n spin system was solved numerically under various conditions. This system is difficult to solve numerically since it involves two time scales: the RF field of 1 kHz, and the relatively slow precession (1 Hz) around the DC field. The accuracy obtained is set by the step size at the 1 kHz level. The results were identical to those obtained with the analytical treatment.

Briefly, in the quantum treatment, effects of static fields both along the RF field (B_0^x, which is a spurious field), and perpendicular to the RF field (applied field $B_0^z \gg B_0^x$), were studied. The unperturbed states are specified by $|\pm 1/2\rangle|n\rangle$ where n is the RF field photon number. The states are degenerate between $\pm 1/2$ before the static fields are applied. Using the formalism developed in [47], the following first-order correction (due to B_0^z) to the $\pm 1/2$ eigenvalues were found:

$$E^{(1)} = \pm\frac{1}{2}\gamma\sqrt{(B_0^x)^2 + [B_0^z J_0(\omega_1/\omega)]^2}, \tag{4.58}$$

where $\omega_1 = \gamma B_{\mathrm{RF}}$. If we require that $E_3^{(1)} = E_{\mathrm{n}}^{(1)}$, the critical dressing condition is obtained. Thus, effects of $B_0^x \neq 0$ enter only in second order. Carrying the perturbation expansion to higher order mixes in states of different n. The second-order corrections are zero, while the third order gives $E^{(3)} \propto (\gamma B_0^z)^3/\omega^2$, which shows that fluctuations in the precession field $\delta\omega_0$ only alter the critical dressing condition to order

$$\delta x_{\mathrm{c}} = (\omega_0/\omega)^2(\delta\omega_0/\omega_0). \tag{4.59}$$

With $\omega_0/\omega \approx 10^4$, the system shows excellent rejection of static field fluctuations.

An important result of this analysis is that the spin/field state cannot be affected by the static electric field. The total system angular momentum could be greater than or equal to one; however, there is no way for the static electric field to couple to the constituent system states (RF photons or particles of spin-1/2).

The issue of motional field systematics was addressed in Sect. 3.5.3, where it was shown that this could be a problem at the level of sensitivity of this proposed experiment. By just considering the effect on the spin precession frequency, the symmetry of the electric field reversal would have to be better than 1%. However, as was shown, the dressing helps alleviate the problem, and relaxes this constraint to around 10%, which is not severe.

Another problem is the pseudomagnetic field, already addressed in relation to the Hg comagnetometer experiment. Again, the pseudomagnetic field is produced by the spin-dependent coherent scattering length contribution to U_F. For ^{3}He at a density of 10^{13} cm^{-3}, the effective magnetic field is 500 nG, which is enormous compared to the anticipated level of accuracy. The problem is that if the $\pi/2$ pulse is imperfect, there will be an effective additional magnetic field on the UCN making the precession frequency time dependent, or variable between fillings. One way to deal with this problem is to reverse the electric field periodically over the course of a measurement cycle. This reduces the overall sensitivity, but effectively eliminates the noise from the pseudomagnetic field. With care, the $\pi/2$ pulse could be accurate to about 10^{-3}, giving a maximum net effective field of 0.5 nG, corresponding to an EDM of 10^{-26} e cm at the anticipated electric field strength. It seems unlikely that the electric field could couple into the system through modification of the pseudomagnetic field; the issues are similar to those outlined in the discussion of the Hg comagnetometer experiment.

As a final note, in the ^{129}Xe experiment to be discussed later, there was a spin shift due to the interaction with the polarized Rb of order 10 μG, a factor of twenty larger than the analogous shift anticipated here. The final experimental limit for the ^{129}Xe experiment was 10^{-26} e cm; given that the electric field was a factor of 10 smaller (thereby making the system more sensitive to an absolute frequency shift) and the spin shift was 20 times larger than the pseudomagnetic field expected for the ^{3}He-UCN experiment, achieving a level of 10^{-28} e cm appears possible. The similarities between the ^{129}Xe experiment and the ^{3}He-UCN experiment proposed here are numerous. The success of the ^{129}Xe shows that complicated magnetometry can be very effective, and illustrates the feasibility of the work proposed here.

4.7 Comparison of Experimental Techniques

Using the figure of merit defined in (3.6),

$$M = ET\sqrt{I},$$

we can compare the various experimental techniques that were discussed in this chapter; the results are summarized in Table 4.3. Of course, each technique has its own systematic problems, but such a comparison is still useful. The systematic common to all experiments is the motional magnetic field; this effect enters into beam experiments as a constraint on the electric and magnetic field alignment, and in UCN storage experiments, as a constraint on the electric field reversal symmetry. In addition, it is assumed that the background magnetic fields can be controlled to the requisite level of accuracy. Also, the effective source intensity I has been extrapolated from that of the original experiments to the current most powerful sources, in those cases where it makes sense to do so.

In Table 4.3, it can be seen that the proposed UCN-^{3}He comparison has promise of orders of magnitude increase in sensitivity over all other experiments.

Table 4.3. Comparison of neutron EDM experimental sensitivities

Technique	E [kV/cm]	T [s]	I [n/s]	Sys. Lim.	$ET\sqrt{I}$
Bragg reflection	1×10^9	2×10^{-7}	10^4	$\theta_{\mathrm{EB}} < 10^{-4}$	2×10^4
Neutron beam magnetic resonance	2×10^5	1.5×10^{-2}	1×10^6	$\theta_{EB} < 10^{-5}$	3×10^6
Ultracold neutron	1×10^4	100	250	$\delta E/E < 0.1$	2×10^7
Pendellösung (α-quartz)	2×10^8	2×10^{-3}	2×10^3	$\theta_{\mathrm{EB}} < 10^{-7}$	2×10^7
UCN-^{3}He	5×10^4	500	2×10^6	$\delta E/E < 0.1$	3.5×10^{10}

5. Theoretical Predictions for Neutron and Electron Dipole Moments

> All the theoretical variants
> are divided into two groups only:
> those in which errors have been
> already found, and those in which
> they still have not.
>
> *V. Sokov*

Just to get a feeling of possible values of the neutron EDM d_n, let us start with a naive estimate as follows:

$$d_n/e \approx m_p^{-1} \times G m_\pi^2 \times \eta \tag{5.1}$$

$$\approx 2 \times 10^{-14}\,\text{cm} \times 2 \times 10^{-7} \times 2 \times 10^{-3}$$

$$\approx 10^{-24} - 10^{-23}\,\text{cm}. \tag{5.2}$$

The first term here, the Compton wavelength of a proton, is a normal scale, that of a nucleon magnetic moment. The second one is the characteristic relative magnitude of weak interactions in hadronic processes (the typical momenta used to construct a dimensionless factor from the Fermi weak interaction constant G are assumed to be comparable to the pion mass m_π). Finally, one may expect that the CP-odd part of the weak interaction is somehow suppressed as compared to the CP-even one. As a natural value of this suppression factor η we take, for the third term, the ratio of the CP-odd and CP-even amplitudes in the decays of K^0 mesons.

This naive estimate is considerably larger than the present experimental upper limit (4.43).

Below we will discuss predictions for the neutron EDM which follow from various models of CP violation. Unfortunately, a detailed discussion of these models demands an acquaintance with elementary particle theory well beyond the level assumed in the present book. Our presentation here will therefore be primarily of a fragmentary style, limited frequently to mere reformulation of theoretical results. Besides original papers cited in the text, the reader is addressed to various textbooks on elementary particle theory, e.g., [150].

Though the experimental limits on the electron EDM will not be discussed until the next chapter, it appears natural to consider the theoretical predictions for it in parallel with those for the neutron EDM (where possible).

Unfortunately, we do not know how to make a simple estimate for the electron dipole moment, analogous to that for the neutron, (5.1). In particular, it is far from clear what should be taken here for the typical momenta to transform G into a dimensionless factor, in other words, what should play the role of m_π. Furthermore, we have no idea and no experimental clue concerning the relative weight of CP-violating leptonic weak interaction as compared to the CP-conserving one; in other words, in this case we have no information about the magnitude of the suppression factor η.

5.1 The CP-Violating θ Term in Quantum Chromodynamics

It is well known that the usual Lagrangian of the electromagnetic field

$$-\frac{1}{4}F_{\mu\nu}F_{\mu\nu} = \frac{1}{2}(\boldsymbol{E}^2 - \boldsymbol{B}^2)$$

can in principle be supplemented by another Lorentz scalar [151]

$$F_{\mu\nu}\tilde{F}_{\mu\nu}, \quad \tilde{F}_{\mu\nu} = \frac{1}{2}\epsilon_{\mu\nu\kappa\lambda}F_{\kappa\lambda}.$$

This scalar violates both P- and T-invariance, which can most easily be seen from its three-dimensional form:

$$F_{\mu\nu}\tilde{F}_{\mu\nu} = -4\boldsymbol{E}\cdot\boldsymbol{B}.$$

However, being a total four-divergence, this scalar generates no observable effects in electrodynamics.

In quantum chromodynamics (QCD) the situation is quite different. Due to the self-interaction of the gluon vector potential A^a_μ, the field configurations which do not fall off rapidly enough at infinity play a prominent role in the theory. Therefore, an analogous four-divergence is no longer harmless. The corresponding possible P- and T-violating term in the QCD Lagrangian is usually written as

$$L_\theta = -\theta\,(\alpha_s/8\pi)\,\tilde{G}^a_{\mu\nu}G^a_{\mu\nu} \tag{5.3}$$

and is called the θ term. Here α_s is the gluon coupling constant, a QCD analogue of the fine structure constant $\alpha = 1/137$ in electrodynamics.

The θ term contribution to the neutron EDM was estimated in [152, 153]. First, by the chiral rotation $\psi \to \exp(-\mathrm{i}\gamma_5\theta)\psi$ of the quark spinor fields ψ the θ term transforms away. However, under this operation the mass term in the Hamiltonian of the light quarks, u, d, s,

$$m_u\bar{u}u + m_d\bar{d}d + m_s\bar{s}s$$

acquires the following CP-odd term

$$\delta H_{CP} = \mathrm{i}\theta\,\frac{m_u m_d m_s}{m_u m_d + m_u m_s + m_d m_s}\,(\bar{u}\gamma_5 u + \bar{d}\gamma_5 d + \bar{s}\gamma_5 s). \tag{5.4}$$

Since among the light quarks the s quark is much heavier than u and d, the mass factor in this expression simplifies:

$$\frac{m_u m_d m_s}{m_u m_d + m_u m_s + m_d m_s} \approx \frac{m_u m_d}{m_u + m_d}.$$

Then, the CP-odd πNN vertex generated by the Hamiltonian (5.4) transforms (with the PCAC technique) as follows:

$$\langle \pi^a N_\mathrm{f} \,|\delta H_{CP}|\, N_\mathrm{i}\rangle = -\theta \frac{m_u m_d}{m_u + m_d} \frac{\sqrt{2}}{f_\pi} \langle N_\mathrm{f} \,|\, \bar{q}\tau^a q \,|\, N_\mathrm{i}\rangle. \tag{5.5}$$

Here τ^a is the isotopic spin operator, $f_\pi = 130$ MeV is the pion decay constant. The nucleon matrix element $\langle p \,|\, \bar{u}d \,|\, n\rangle$ is related by ordinary SU(3) symmetry to the mass splitting in the baryon octet:

$$\langle p \,|\, \bar{u}d \,|\, n\rangle = \bar{p}n \frac{M_\Xi - M_\Sigma}{m_s} \approx \bar{p}n. \tag{5.6}$$

Here p and n are the Dirac spinors of the proton and neutron; M_Ξ and M_Σ are the masses of the Ξ and Σ hyperons, respectively. The full πNN interaction can be described now by the effective Hamiltonian

$$H_{\pi NN} = \boldsymbol{\pi} \bar{N} \boldsymbol{\tau} \,(\mathrm{i}\gamma_5 g_{\pi NN} + \bar{g}_{\pi NN})\, N, \tag{5.7}$$

where the CP-odd constant is

$$\bar{g}_{\pi NN} = -\theta \frac{m_u m_d}{m_u + m_d} \frac{\sqrt{2}}{f_\pi} \frac{M_\Xi - M_\Sigma}{m_s} \approx -0.027\theta. \tag{5.8}$$

The CP-even πNN constant in the effective Hamiltonian is well known:

$$g_{\pi NN} = 13.6. \tag{5.9}$$

The crucial observation made in [153] is that in the chiral limit $m_\pi \to 0$ the neutron EDM can be expressed rigorously via $\bar{g}_{\pi NN}$ and $g_{\pi NN}$. In this limit there are only two diagrams, presented in Figs. 5.1a,b, which are singular in the pion mass. Here, one πNN vertex is the strong pseudoscalar coupling, with the coupling constant $g_{\pi NN}\sqrt{2}$, and the second, a scalar one, is CP-odd, with the coupling constant $\bar{g}_{\pi NN}\sqrt{2}$. The contribution of these diagrams to the neutron EDM is

$$d_\mathrm{n} = \frac{e}{m_\mathrm{p}} \frac{g_{\pi NN}\, \bar{g}_{\pi NN}}{4\pi^2} \ln \frac{m_\rho}{m_\pi}. \tag{5.10}$$

The choice of the ρ meson mass $m_\rho = 770$ MeV as the typical hadronic scale at which the logarithmic integral is cut off at high momenta, is of course to some degree a matter of convention.

In fact, the chiral parameter, the logarithm in expression (5.10), at any reasonable cut-off is not large; it is only 1.7 with our choice of m_ρ for the cut-off. However, due to the absence of other terms logarithmic in m_π, strong mutual cancellation between this contribution and others seems quite unnatural. In this sense expression (5.10) can be considered a rather conservative estimate of the neutron EDM.

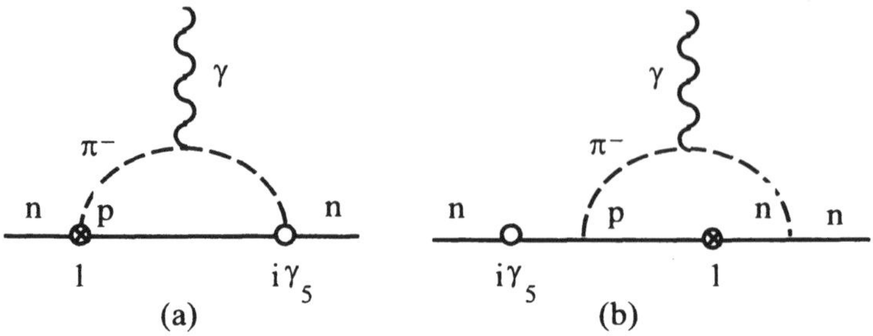

Fig. 5.1a,b. Chiral contribution to the neutron EDM

Combining this estimate with the experimental result (4.43), we get a very strict upper limit for the CP-odd QCD parameter

$$|\theta| < 3 \times 10^{-10}. \tag{5.11}$$

The same result can be formulated as an upper limit on the effective CP-odd constant for a charged pion:

$$\bar{g}_{\pi NN} < 7 \times 10^{-12}. \tag{5.12}$$

A word of caution is warranted here. The whole problem of the θ term seems rather mysterious. There is no natural explanation for the extreme smallness of the parameter θ. It is even more unnatural when one considers that this parameter is renormalized by other CP-odd interactions, and in general its renormalization can be infinite. In particular, in the standard model the induced contributions to θ may diverge logarithmically starting from 14th(!) order in the electroweak coupling constant [154, 155]. This means in fact that the parameter θ, strictly speaking, cannot be calculated.

One of the solutions of this so-called strong CP problem would be to leave the θ term as it is, but to somehow make it harmless. This could be achieved by introducing an extra global symmetry into the theory [156]. However, such a symmetry leads in turn to the prediction of a new light pseudoscalar particle, known as the axion [157, 158]. But numerous experimental searches for the axion give no indication of its existence. Another way out, which is quite obvious from formula (5.4), is to assume that one of the quark masses is zero. This assumption, however, certainly contradicts experimental data.

In view of these uncertainties and difficulties, the common approach to the θ term problem as related to the neutron EDM is as follows. The phenomenological upper limit (5.11) is taken as it is. But the contributions to the neutron EDM induced by radiative corrections to the θ term are not considered, though in some models they are formally far from being negligible.

To conclude this section, let us mention one more purely gluonic CP-odd operator, cubic in the field strength $G^a_{\mu\nu}$, which is considered in [159]. As distinct from the θ term, its coefficient is calculable in terms of the weak parameters, and this operator certainly contributes to the neutron dipole moment. However, in no model explored does this contribution dominate the EDM [160].

5.2 Predictions of the Standard Model for Dipole Moments

The only source of CP violation in the standard model is the phase δ in the Kobayashi–Maskawa (KM) matrix of flavor mixing in the charged current interaction [161]:

$$(\bar{u}, \bar{c}, \bar{t}) \begin{pmatrix} c_1 & s_1 c_3 & s_1 s_3 \\ -s_1 c_2 & c_1 c_2 c_3 - e^{i\delta} s_2 s_3 & c_1 c_2 s_3 + e^{i\delta} s_2 c_3 \\ s_1 s_2 & -c_1 s_2 c_3 - e^{i\delta} c_2 s_3 & -c_1 s_2 s_3 + e^{i\delta} c_2 c_3 \end{pmatrix} \begin{pmatrix} d \\ s \\ b \end{pmatrix} . \quad (5.13)$$

Here $c_i = \cos\theta_i$, $s_i = \sin\theta_i$, $\theta_i (i = 1, 2, 3)$ are mixing angles.

Obviously, an electric dipole moment (EDM) cannot arise in the standard model to first order in G [162–164]: any element of the KM matrix (5.13) enters into a single W boson exchange, which is diagonal in flavor, via its modulus squared; this cancels out the CP violation. Therefore, in the context of the standard model, an EDM appears as a second-order effect. Still, since the theory is renormalizable, an EDM could be of order αG. But due to strong cancellations between the contributions of different virtual quarks [165], the true order of magnitude is at least G^2. Moreover, it was demonstrated [166] that a quark does not acquire an EDM even in this order. The same is true for the W boson dipole moment [167]. It should be noted that in some sense the last EDM cancelation is accidental; another CP-odd electromagnetic moment of the W boson, the magnetic quadrupole, is induced in the two-loop, G^2 approximation [168].

The electron EDM cannot originate in the standard model even from three-loop diagrams [167]; the prediction from the standard model for the electron EDM is tiny:

$$d_e < 10^{-40}\, e\,\text{cm}. \quad (5.14)$$

The quark EDM is generated in the standard model only at order $\alpha_s G^2$, which is extremely small, $10^{-34} - 10^{-33}$ $e\,$cm [169]. The neutron dipole moment is controlled in the standard model by long-distance contributions, and can be estimated in the same chiral limit. The diagrams logarithmic in m_π are dominated numerically by those presented in Fig. 5.2. Here the symbol $\circ$ refers to the usual S wave nonleptonic hyperon decay amplitude A. The CP-odd pseudoscalar vertex $\otimes$ is generated by the so-called "penguin" diagram. This contribution to the neutron EDM is [170]

$$d_{\rm n} = -\,e\,G^2 \sin\delta\, s_1 s_2 s_3 c_2 \frac{\alpha_s}{\sqrt{2}\,27\,\pi^3} \ln\frac{m_t^2}{m_c^2}$$

$$\times \frac{f_\pi\, m_\pi^4}{m_s\,(m_u + m_d)}\, A\,(2\alpha - 1)\, g_A \ln\frac{m_K}{m_\pi}. \tag{5.15}$$

Here m_t, m_c , and m_K are the masses of the t quark, c quark, and K meson, respectively; $\alpha \approx 0.64$ is the relative weight of the D coupling in the interaction of the baryon and meson octets; $g_A = 1.26$. Numerically this estimate is

$$d_{\rm n} \approx 10^{-32} - 10^{-31}\; e\,{\rm cm}. \tag{5.16}$$

An estimate for the neutron EDM one order of magnitude larger was given in [171].

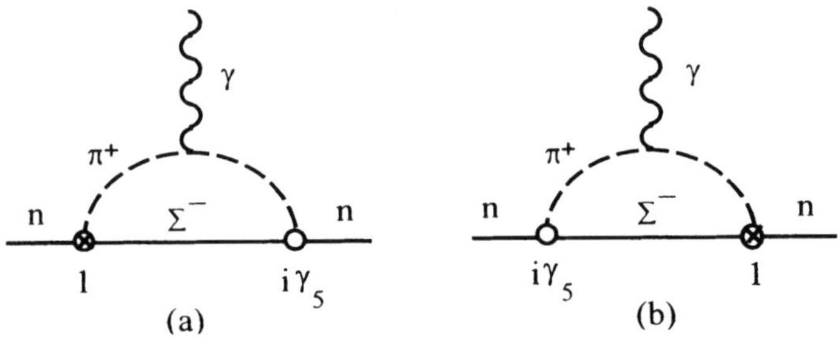

Fig. 5.2a,b. Long-distance contribution to the neutron EDM in the standard model

Let us mention here that the θ term is induced in the standard model at the same $\alpha_s\, G^2$ order (four-loop approximation) [169]. Being generated in this way, it is

$$\theta \approx 10^{-19} - 10^{-18}. \tag{5.17}$$

It was pointed out recently that a minimum extension of the standard model to four generations with heavy new quarks can raise the magnitude of the theoretical prediction for the light quark EDM, in the same $\alpha_s\, G^2$ approximation, to $3 \times 10^{-32}\; e\,{\rm cm}$ [172]. Moreover, if the quarks of the fourth generation are as heavy as 500 GeV, the three-loop diagrams with Z boson exchange, instead of gluon exchange, dominate and bring the prediction for the neutron EDM to $10^{-29}\; e\,{\rm cm}$ [173].

5.3 Spontaneous *CP* Violation in the Higgs Sector

The possibility of CP violation being generated by the spontaneous symmetry breaking in the Higgs fields interaction was pointed out in [174]. A more realistic model based on this idea was suggested later [175] and contains at least three doublets of complex Higgs fields.

In the most ambitious approach one may try to ascribe to this mechanism the CP-odd effects observed in K meson decays [176, 177]. This approach has a shortcoming in that it requires the Higgs boson masses to be uncomfortably small. Moreover, accurate estimates for the neutron EDM in this version [170,178–180] lead to predictions

$$d_{\mathrm{n}} \approx 10^{-24} - 10^{-23}\, e\,\mathrm{cm}, \tag{5.18}$$

which are much greater than the experimental upper limits.

But then one can pass over to a more "natural" version of this model, with heavy Higgs bosons. Of course, in this case the model is responsible for only a small portion of CP violation in kaon decays. It would be new physics, a new source of CP violation, supplemental to that generating the effects already observed. Even if the Higgs bosons are heavy, the predictions of the model for the dipole moments of both neutron and electron are, at least, very close to the present upper limits. The dominant contribution to the dipole moments is given by diagrams of the type shown in Fig. 5.3 with a heavy particle (t quark, W boson or Higgs) propagating in the upper loop [181]. Since the Higgs coupling to a particle (Higgs itself included) is proportional to the particle mass, we gain more by having Higgs couple to a heavy particle in the upper loop than we loose by going over to a higher order of perturbation theory. The corresponding prediction for the electron EDM is [181–183]

$$d_{\mathrm{e}} \approx 10^{-27}\, e\,\mathrm{cm}. \tag{5.19}$$

For the neutron dipole moment this approach is further elaborated upon in [184–186]. In particular, it is pointed out there that, in the model discussed,

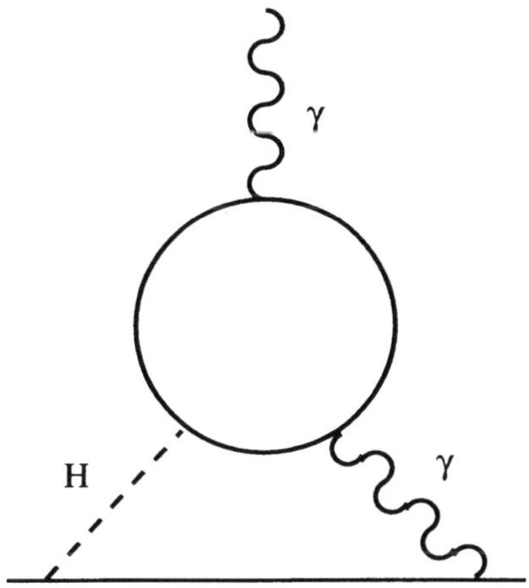

Fig. 5.3. Two-loop contribution to an electric dipole moment

the neutron EDM is controlled by the diagram given in Fig. 5.4 with the t quark propagating in the upper loop and both wavy lines corresponding to gluons. The effective operator generated by this diagram is

$$H_{qg} = \frac{d^c(q)}{2}\,\bar{q}\gamma_5\sigma_{\mu\nu}t^a q\,G^a_{\mu\nu}, \tag{5.20}$$

where $t^a = \lambda^a/2$ are the generators of the color SU(3) group. From comparison with expression (2.5) it is only natural to call the constant $d^c(q)$ in (5.20) the quark chromoelectric dipole moment (CEDM). Its value, as obtained directly from the diagram of Fig. 5.4, refers in fact to a high momentum scale, of order 100 GeV. The renormalization group factor necessary to bring $d^c(q)$ down to the common hadronic scale, is well known [184,185,187]. The resulting estimate for the quark CEDM is

$$d^c(q) \approx 3 \times 10^{-25}\ \text{cm}. \tag{5.21}$$

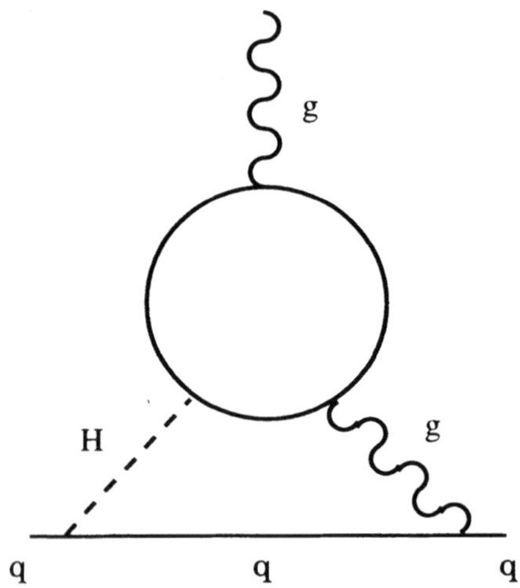

Fig. 5.4. Two-loop contribution to the quark CEDM

However, the most serious problem is to find the CEDM contribution to the neutron dipole moment. Here our predictions somewhat exceed those of [184, 185]. The simplest way to estimate this contribution is to assume, for purely dimensional reasons, that d_n/e is roughly equal to $d^c(q)$ [188–190] (obviously, the electric charge e should be removed from d_n, being a parameter unrelated to the nucleon structure).

In a more elaborate approach [190], the neutron EDM is estimated in the chiral limit via formula (5.10). For both u and d quarks, the contribution of operator (5.20) to the CP-odd πNN constant $\bar{g}_{\pi NN}$ is transformed by the PCAC technique to the same expression:

$$\langle \pi^- p \,|\, g\,\bar{q}\gamma_5\sigma_{\mu\nu}t^a q\, G^a_{\mu\nu} |\, n\rangle = \frac{\mathrm{i}}{f_\pi} \langle p \,|\, g\,\bar{u}\sigma_{\mu\nu}t^a d\, G^a_{\mu\nu}\,|\, n\rangle. \tag{5.22}$$

We include the quark–gluon coupling constant g explicitly in the above relation since the corresponding estimate based on the QCD sum rules refers directly to the last matrix element. This estimate gives a value close to -1.5 GeV^2. For momenta ≈ 1 GeV in this estimate, we take $g \approx 2$. Then the result for the neutron EDM is [191]:

$$d_{\mathrm{n}} \approx 2 \times 10^{-25}\, e\,\mathrm{cm}. \tag{5.23}$$

Let us introduce the ratio of the neutron dipole moment, as induced by a CEDM, to $d^{\mathrm{c}}(q)$ itself:

$$\rho = \frac{d_{\mathrm{n}}/e}{d^{\mathrm{c}}(q)}. \tag{5.24}$$

Its value obtained in this approach, $\rho = 0.7$, is indeed quite close to unity. In our opinion, this good agreement with the above simple result indicates the reliability of both estimates.

An essential contribution to the neutron EDM can stem from the chromoelectric dipole moment $d^{\mathrm{c}}(s)$ of the s quark [186]. The large magnitude of $d^{\mathrm{c}}(s)$, as compared to $d^{\mathrm{c}}(d)$, results from the large ratio of the quark masses, $m_s/m_d \approx 20$.

On the other hand, for the s quark, the ratio

$$\rho_s = \frac{d_{\mathrm{n}}/e}{d^{\mathrm{c}}(s)} \tag{5.25}$$

should be much smaller than unity. Indeed, according to the QCD sum rule calculations of [180], it is about 0.1. One should mention that other estimates [178, 192] predict the ratio (5.25) to be an order of magnitude smaller, and this smaller value was used in [186].

How reliable, then, is the estimate $\rho_s = 0.1$? There are strong indications now that the admixture of the $\bar{s}s$ pairs in nucleons is quite considerable. In particular, this relates to the spin content of a nucleon. And though these indications refer to operators different from $\bar{s}\gamma_5\sigma_{\mu\nu}\, t^a s\, G^a_{\mu\nu}$, they give serious reasons to believe that the estimate

$$\rho_s = 0.1 \tag{5.26}$$

is a conservative one.

At this value of ρ_s the resulting contribution of the s quark CEDM to the neutron dipole moment [191]

$$d_{\mathrm{n}} = 6 \times 10^{-25}\, e\,\mathrm{cm} \tag{5.27}$$

is significantly larger than the experimental upper limit (4.43).

Thus, very special assumptions concerning the parameters of the model of spontaneous CP violation in the Higgs sector (such as the large mass of the Higgs boson, small values of the CP-violating parameters, etc.) are necessary to reconcile the predictions of this model with the experimental upper limits for the neutron electric dipole moment. But with such a fine-tuning, the prediction (5.19) for the electron EDM decreases as well.

Similar conclusions can be drawn concerning CP violation in the supersymmetric SO(10) model of grand unification. The neutron dipole moment, as induced by the quark EDM, is calculated in [193]. Its typical value turned out to be somewhat smaller than the experimental upper limit. The electron EDM at the same representative values of the parameters of the model is very close to the experimental bound. However, here too the s quark CEDM is much more efficient in inducing the neutron dipole moment. At the natural values of the parameters of the model, this contribution to the neutron EDM is [194]

$$d_{\mathrm{n}} = 4 \times 10^{-25}\, e\,\mathrm{cm}, \tag{5.28}$$

which again exceeds appreciably the experimental upper limit.

5.4 Phenomenological Approach

As mentioned already, the neutron EDM experiments have ruled out a large number of theoretical models of CP violation. But the surviving models are still too numerous to discuss them all in the present book. Moreover, most of these models contain too many CP-violating parameters; examples include supersymmetric grand unified theories and left–right symmetric models. In particular, the last class of models contains, generally speaking, in line with the usual CP-odd Kobayashi–Maskawa phase for left quarks, an analogous phase for right quarks, as well as a CP-violating mixing parameter for the gauge bosons.

Since various models of CP violation have, as a rule, too many degrees of freedom, in the present section we will deviate from the main trend of this chapter. Instead of theoretical *predictions* for the neutron EDM, we present here the implications of the experimental result (4.43) in a purely phenomenological way. Following [190, 195], we construct CP-odd quark–quark, quark–gluon and gluon–gluon operators of low dimension, and find upper limits on the corresponding coupling constants.

Let us write the CP-odd Hamiltonian of quark–quark interaction as

$$\begin{aligned} H &= \frac{G}{\sqrt{2}} \sum_i k_i O_i, \\ O_s &= (\bar{q}_1 \mathrm{i}\gamma_5 q_1)(\bar{q}_2 q_2), \\ O_s^c &= (\bar{q}_1 \mathrm{i}\gamma_5 t^a q_1)(\bar{q}_2 t^a q_2), \\ O_t &= \frac{1}{2}\, \epsilon_{\mu\nu\alpha\beta} (\bar{u}\sigma_{\mu\nu} u)\, (\bar{d}\sigma_{\alpha\beta} d), \end{aligned} \tag{5.29}$$

$$O_t^c = \frac{1}{2}\,\epsilon_{\mu\nu\alpha\beta}(\bar{u}\sigma_{\mu\nu}t^a u)\,(\bar{d}\sigma_{\alpha\beta}t^a d).$$

In this set of equations, the summation given in the first line is over the operators given in the next lines. The operators O_s and O_s^c refer to the cases of the same quarks ($q_1 = q_2 = u, d$)) and different ones ($q_1 \neq q_2$). As to CP-odd tensor operators without and with t^a, at $q_1 = q_2$ they reduce via the Fierz transformation to O_s and O_s^c. The results will be formulated as upper limits on the dimensionless constants k_i characterizing the magnitude of a corresponding interaction in the units of the Fermi constant G.

The CP-odd vertices of interest to us transform by means of the PCAC technique as follows:

$$\langle \pi^- p\,|\,O_i\,|\,n\rangle = \frac{\mathrm{i}}{f_\pi}\,\langle p\,|\,[O_i,\,u^\dagger\gamma_5 d]\,|\,n\rangle. \tag{5.30}$$

To estimate the arising matrix elements we reduce them by the vacuum factorization to

$$\langle p\,|\,[O_i,\,u^\dagger\gamma_5 d]\,|\,n\rangle = \mathrm{i}\,\langle\bar{q}q\,\rangle\,\langle\, p\,|\,\bar{u}d\,|\,n\rangle\, C_i. \tag{5.31}$$

The numerical factor C_i here depends on the specific form of the operator O_i:

$$C_s = \frac{5}{6},\quad C_s^c = -\frac{2}{9},\quad C_t = 2,\quad C_t^c = \frac{8}{3}.$$

For the vacuum condensate we assume the numerical value

$$\langle\bar{q}q\rangle = -\,(0.25\,\mathrm{GeV})^3.$$

Estimate (5.6) is used for the last nucleon matrix element. Finally, using the upper limit (5.12) we get the following results [190]:

$$|\,k_s|\, < \,3\times 10^{-5},\ |\,k_s^c|\, < \,9\times 10^{-5},\ |\,k_t|\, < \,10^{-5},\ |\,k_t^c|\, < \,7\times 10^{-6}. \tag{5.32}$$

The derivation of the upper limits on the CEDM $d^c(q)$ of the light, u and d quarks, as well as on $d^c(s)$, is outlined in fact in the previous section.

Recently, upper limits on some CP-odd quark–quark and quark–electron interaction constants were obtained by calculating directly quark and electron EDMs in the one-loop approximation [196]. All those limits on CP-odd interactions of light quarks are substantially weaker than (5.32). As to the constraints from [196] on the quark–electron interactions, much better upper limits follow for them in a different way, from atomic experiments (see Sect. 8.3).

As to the pure gluon CP-odd operators, the θ term has been discussed already, together with its upper limits, in Sect. 5.1. The next operator, cubic in gluon field strength, which was only mentioned in that section, can be written as

$$\frac{G}{\sqrt{2}}\,k^g\,\frac{1}{6}\,\epsilon_{\mu\nu\alpha\beta}f^{abc}G^a_{\mu\nu}G^b_{\alpha\rho}G^c_{\beta\rho}. \tag{5.33}$$

Here f^{abc} are the structure constants of the SU(3) group. The upper limit from the neutron experiments for the dimensionless constant k^g of this operator is [197]

$$|k^g| < 4 \times 10^{-5}. \tag{5.34}$$

Finally, the upper limit (6.5) for the electron EDM constrains the CP-odd electron–electron interaction

$$H_e = \frac{G}{\sqrt{2}} k_s^e (\bar{e}\, \mathrm{i}\gamma_5 e)(\bar{e}\, e). \tag{5.35}$$

Its contribution to the electron EDM can be calculated in the spirit of [196]:

$$d_{\mathrm{e}} = e \frac{G}{\sqrt{2}} \frac{k_s^e m_{\mathrm{e}}}{8\pi^2} \log \frac{\Lambda^2}{m_{\mathrm{e}}^2}. \tag{5.36}$$

In this way, at the cut-off parameter value $\Lambda \approx 100$ GeV, we obtain [198]

$$k_s^e < 2 \times 10^{-4}. \tag{5.37}$$

The limits on the effective CP-odd interaction operators presented above are quite impressive; all the constants are several orders of magnitude smaller than the usual Fermi weak interaction constant G.

6. EDM Experiments with Paramagnetic Atoms

> These things must be done delicately,
> Or you hurt the spell...
> *The Wicked Witch of the West*

The problem of measuring an electric dipole moment among the constituents of an electrostatically bound system (atom) was addressed with the first mention of EDMs in the paper of Purcell and Ramsey [4]; an intuitive explanation of the shielding effect was first explicitly formulated in [199]. This problem was briefly lost sight of in Fairbank's proposal to measure a ^{3}He EDM in a cryogenic system, but led Schiff to write his famous paper; the shielding effect is sometimes referred to as Schiff's theorem, but in fact the purpose of [200] was to illustrate ways around the shielding problem; if the system constituents are of finite size or if there are other (magnetic) forces, the shielding argument does not apply.

6.1 The Shielding Problem

The shielding effect is illustrated as follows. For a neutral atom of non-relativistic point particles, in equilibrium under the action of electrostatic forces, there is no effect due to the EDMs of the constituent particles [199]. This is because the atom remains at rest when an external electric field $\boldsymbol{E}_{\text{ext}}$ is applied; the implication is that the total electric field acting on each particle must vanish. This comes about by an internal rearrangement of the system's constituents giving rise to an internal field $\boldsymbol{E}_{\text{int}}$ that exactly cancels $\boldsymbol{E}_{\text{ext}}$ at each charged particle, as required by the static equilibrium condition. Thus, there is no observable effect due to dipole moments of the system's constituents; the external field is effectively switched off.

A quantum mechanical proof of this statement is given in [200] and proceeds as follows. Consider a system comprising particles with charge e_k and dipole moment $\boldsymbol{d}_k$. In addition to a "bare" EDM given by

$$\sum_k \boldsymbol{d}_k,$$

there is an EDM induced by the mixing of states of opposite parity through the interaction

$$H_{\mathrm{d}} = -\sum_k \boldsymbol{d}_k \cdot \boldsymbol{E}_{\mathrm{int}}(\boldsymbol{r}_k), \tag{6.1}$$

which can be transformed to

$$H_{\mathrm{d}} = \sum_k \frac{1}{e_k} \boldsymbol{d}_k \cdot \boldsymbol{\nabla}_k U(\boldsymbol{r}) = \mathrm{i} \sum_k \frac{1}{e_k} [\boldsymbol{p}_k, H_0], \tag{6.2}$$

where U and H_0 are the potential energy and Hamiltonian of the unperturbed system (i.e., without dipole moments) and it is assumed that the interparticle interactions are purely electrostatic. The mixed state resulting from this interaction is

$$\widetilde{|n\rangle} = |n\rangle + \sum_m \frac{|m\rangle\langle m|H_{\mathrm{d}}|n\rangle}{E_n - E_m} = \left(1 + \mathrm{i}\sum_k \frac{1}{e_k} \boldsymbol{d}_k \cdot \boldsymbol{p}_k\right)|n\rangle.$$

The dipole moment of this mixed state is

$$\langle n| \left(1 - \mathrm{i}\sum_k \frac{1}{e_k} \boldsymbol{d}_k \cdot \boldsymbol{p}_k\right) \sum_l e_l \boldsymbol{r}_l \left(1 + \mathrm{i}\sum_m \frac{1}{e_m} \boldsymbol{d}_m \cdot \boldsymbol{p}_m\right)|n\rangle$$

$$= \langle n|\mathrm{i}\left[\sum_l e_l \boldsymbol{r}_l, \sum_k \frac{1}{e_k} \boldsymbol{d}_k \cdot \boldsymbol{p}_k\right]|n\rangle = -\langle n|\sum_k \boldsymbol{d}_k|n\rangle,$$

which exactly cancels the "bare" EDM.

The EDM of the nucleus remains completely shielded even if relativistic effects in the electron motion are included. This is because (6.2) remains valid even though the form of the Hamiltonian changes. For the same reason, the nuclear EDM remains shielded when interactions leading to the Lamb shift are taken into account [201].

Obviously, it is impossible to apply an electric field to free electrons or protons because they would simply be accelerated out of the measuring apparatus. Measurements on atoms appeared hopeless because of the shielding problem; hence the choice of the neutron for the first experiment. Nonetheless, limits on the proton and electron EDMs were extracted from an analysis of the hydrogen spectrum [4, 202]. As a related question, an ion contained in a radio-frequency quadrupole trap (Paul trap) will be effectively shielded from external static electric fields; on application of a static field, the ion moves to a new point of equilibrium. This implies that the average static electric field at the average position of the ion is zero.

The electron EDM limit was improved, oddly enough with free electrons, by searching for a motional *electric* field in connection with an experimental measurement of the electron g factor [203]. The idea is that in a frame moving in a magnetic field, a motional electric field $\boldsymbol{E} = \boldsymbol{B} \times \boldsymbol{v}$ appears, and this field can couple to the internal spin states of the free electron through the $\boldsymbol{d}_{\mathrm{e}} \cdot \boldsymbol{E}$ interaction. The electron spin was initially perpendicular to, and precessing in, a static magnetic field $\boldsymbol{B}$, and if there was an EDM interaction, the spin would also precess around $\boldsymbol{E}$, eventually giving rise to a spin component along $\boldsymbol{B}$. The limit obtained was about three orders of magnitude more strict than that obtained from the hydrogen spectrum.

In spite of the apparently hopeless situation in regard to atomic EDM measurements, two experiments on alkali atoms were performed. The first atomic experiment, done in 1962 in Seattle, used an optically pumped Rb magnetometer with an internally applied electric field [204]. The second experiment, done in 1964 at Brandeis, used a Cs beam [205]; this experiment was noteworthy in that it included the first study of the $\boldsymbol{v} \times \boldsymbol{E}$ field [74]. This experiment was not devised as a serious test of time-reversal symmetry, but as a simple test of the new apparatus [206]; at the time, all evidence indicated a clear distinction between parity, which was known to be violated, and time reversal, apparently a good symmetry. It was during the time of this experiment that CP violation was discovered in K^0 decay, thereby changing the situation. These experiments immediately became of great interest, and Sandars did the calculations to determine the degree of shielding within the alkali atoms; a startling result was obtained, as described in the next section.

6.2 Enhancement of the Electron EDM in Paramagnetic Atoms

An electron EDM can induce a net atomic EDM because magnetic forces, in addition to electrostatic forces, are important. The magnetic interactions increase rapidly with Z, the atomic number. As was discovered by Sandars, the net EDM of a heavy atom can in fact be many times larger than the electron EDM [207]. This discovery provided the impetus for the experimental search for EDMs of heavy atoms.

The magnitude of the enhancement can be simply estimated [23, 208]. Opposite-parity states are admixed because of the interaction of the electron EDM with the internal atomic field $\boldsymbol{E}_{\text{int}}(\boldsymbol{r})$; making allowances for the relativistic nature of the effect, its order of magnitude is

$$K \sim ea\langle E_{\text{int}} v^2 \rangle / \text{Ry},$$

giving an induced dipole of order

$$ead_{\text{e}}\langle E_{\text{int}} \rangle / \text{Ry} = Kd_{\text{e}},$$

where K is the enhancement factor. The interaction which gives rise to the mixing of opposite-parity levels is sufficiently singular for $r \to 0$ so that the main contribution to K occurs at short distances, $r \sim a/Z$. In this region, of volume $\sim \pi a^3/Z^3$, the internal electric field is of order

$$E_{\text{int}} \sim Ze/r^2 \sim Z^3 e/a^2,$$

while

$$v^2 \sim Z^2\alpha^2$$

and, according to §71 of Ref. [15],

$$|\psi|^2 \sim Z/\pi a^3.$$

Taking these factors together, along with the typical atomic energy interval $\sim e^2/a$, gives

$$K \sim ea\frac{Z^3 e}{a^2} Z^2\alpha^2 \frac{Z}{\pi a^3}\frac{\pi a^3}{Z^3}\frac{a}{e^2} \sim Z^3\alpha^2, \tag{6.3}$$

which is much greater than unity in heavy atoms. Table 6.1 gives a tabulation of calculated enhancement factors; the Z^3 dependence is evident. In fact, the dependence is slightly larger than Z^3; this is the result of a relativistic enhancement of the wavefunction near the nucleus (see Sect. 9.1).

Table 6.1. Calculated electron EDM enhancement factors for paramagnetic atoms

Atom	Z	Enhancement factor K	$K/(\alpha^2 Z^3)$	Reference
Rb	37	24	8.9	[209]
Xe($^3P_2$)	54	120	14.3	[208, 210]
Cs	55	114	12.9	[211]
Tl	81	−585	−20.7	[212]
Fr	87	1150	32.8	

6.3 Overview of Paramagnetic Atom Experiments

Table 6.2 lists some of primarily paramagnetic atom experiments which have yielded limits for the electron EDM. The evolution of the electron EDM limit follows an approximately two-order-of-magnitude decrease per decade, similar to that for the neutron EDM (Fig. 4.1).

Before discussing the two most sensitive experiments in detail, we will review the interesting features of some of the experiments listed in this table. First, the Rb experiment of 1962 (the results of which were reported in 1964) made use of optical pumping techniques with the atoms contained in a vapor cell [204]. As was pointed out, a significant advantage of the storage cell technique is that the motional field effect is largely eliminated because the average velocity of the atoms is zero.

In the electron–^{12}C scattering experiment [213], a rise in the cross section as $\theta \to 180°$ would be evidence for an electron EDM, and failure to observe such an effect led to the above limit.

To eliminate the motional field effect, the Cs-beam experiment of 1968 had provisions to send Na or K through the same region [214]. The idea was

Table 6.2. Experimental limits on paramagnetic atom EDMs and inferred limit for the electron EDM

Atom or System	Date	Reference	$d(a)$ [e cm]	d_e [e cm]
H $2s$-$2p$ (Lamb shift)	1950,1958	[4, 202]		2×10^{-13}
Electron $g-2$	1959	[203]		4×10^{-16}
Rb $5s$	1962	[204]	1×10^{-18}	5×10^{-20}
Cs $6s$	1964	[205]	2×10^{-19}	1×10^{-20}
$e-{}^{12}$C scattering	1965	[213]		2×10^{-16}
Cs $6s$	1968	[214]	3×10^{-22}	3×10^{-24}
Tl $6p_{1/2}$	1970	[215]	5×10^{-21}	1×10^{-23}
Xe $6s6p\,^3$P$_2$ metastable	1970	[210]	4×10^{-22}	2×10^{-24}
Fe^{+3} $d_{3/2}$	1978	[216]		2×10^{-22}
Rb $5s$	1987	[217]*	1.2×10^{-23}	5×10^{-25}
Cs $6s$	1989	[218]	1.3×10^{-23}	1×10^{-25}
Tl $6p_{1/2}$	1990	[219]	1×10^{-23}	2×10^{-26}
Tl $6p_{1/2}$	1994	[220]	2×10^{-24}	4×10^{-27}

* Unpublished result

that as the angle between $\boldsymbol{E}$ and $\boldsymbol{B}$ was varied, the angle where the effect of the electric field (through the motional field and a possible true EDM interaction) was zero would be different for the two species if there was a Cs EDM. The 1970 Tl $p_{1/2}$ (ground state) experiment was performed with an atomic beam and also had provisions for sending alkali atoms through the apparatus [215].

A different method of minimizing the motional field was used in the metastable Xe experiment of 1970 [210]. This state has a large electrical polarizability, and to extract an EDM by measuring the spin precession frequency in the usual way would require an electric field reversal symmetry of 1 part in 10^{10}. The trick employed to get around this was to measure the frequency by a multiple quantum transition so that $\Delta|m_J| = 0$, for which there is no polarizability contribution. On the other hand, the large polarizability meant that the electric field defined the quantization axis; the $\boldsymbol{B}_\mathrm{m} = \boldsymbol{v}\times\boldsymbol{E}/c$ motional field is therefore perpendicular to the quantization axis identically, giving a reduction in motional field effects by a factor $1/|\boldsymbol{E}|^3$. This dependence results because the motional field amplitude is proportional to $|\boldsymbol{E}|$, and gives a correction only in second order because the motional field is perpendicular to the quantization axis. If there are magnetic field components both parallel and perpendicular to the electric field, in the limit where $B_\perp \ll B_m \ll B_\|$,

$$\Delta f \propto \frac{B_\mathrm{m} B_\perp B_\|}{E^4}.$$

The Fe^{3+} experiment is worthy of mention in that it employed a completely different technique, originally suggested by F.L. Shapiro [117]. The idea was to apply a strong electric field to a material which contains unpaired electron spins, and then look for a change in the sample magnetic flux. The material used, a nickel-zinc ferrite, has a high magnetic susceptibility and low conductivity, along with a high concentration of unpaired electrons. An electric field of up to 2 kV/cm was applied to the sample which was held at liquid helium temperature, and the magnetization change was determined using a SQUID magnetometer [216].

The result of the 1987 Rb experiment was never published because of some concerns over systematic effects [217]. Nonetheless, this work led to the study of a linear electric field effect on the atomic absorptivity [221, 222]. The details of the experiment are in many ways similar to those of the Cs work described below.

6.4 The Cs EDM Experiment

The most recent work on Cs (enhancement factor of about 120) has been carried out at Amherst College; the last reported result was in 1989, with [218]

$$d(\mathrm{Cs}) = (2.3 \pm 7.8 \pm 1.4) \times 10^{-24}\, e\,\mathrm{cm} \longrightarrow d_\mathrm{e} < 10^{-25}\, e\,\mathrm{cm}, \tag{6.4}$$

where first error is statistical and reflects subtraction of possible systematic errors, and the second is a possible systematic due to leakage currents. It should be noted that the system shot noise (photon counting) is about three orders of magnitude less than the systematic error. At the time of publication, this result represented a leap by nearly four orders of magnitude in the experimental limit for the electron EDM.

With improvements to the apparatus, and additional data acquisition, a factor of 20 improvement in the statistical accuracy has been obtained. Unfortunately, non-statistical fluctuations between differing configurations of the apparatus have limited the net improvement to about a factor of 5 over the previous result.

This experiment employed a stack of four cells, with the outer two serving as a magnetometer, and the inner two having oppositely directed electric fields. Electric fields of up to 4 kV/cm were applied, with resulting leakage currents across the cells of less than 20 pA each. If this current flowed in a half-loop around one cell, a false Cs EDM of $1.4 \times 10^{-24}\, e\,\mathrm{cm}$ could result, and this was taken as the second systematic error [see (3.37) in Sect. 3.5.1].

The cells were constructed of square cross section glass tubing (2.5 cm^2 dimension), with tin-oxide coated glass electrodes glued to the top and bottom with vacuum epoxy. To improve the lifetime and high voltage characteristics, the cells contained 200 torr of N_2 in addition to the small amount of Cs metal.

The spin lifetime was on the order of 20 ms, limited by disorienting collisions with the N_2.

The experiment was operated with zero applied magnetic field, as described in Sect. 3.2.1 [see (3.37)]. Magnetic shields and internal compensation coils were used to reduce the net average field at each cell to less than 100 nG.

Semiconductor diode lasers were used to optically pump the ground state; circularly polarized 894 nm radiation, tuned to the $6s_{1/2}$, $F = 3 \rightarrow 6p_{1/2}$ transition, was used. (Optical pumping via the $6p_{1/2}$ is much more efficient than with the $6p_{3/2}$ excited state.) Spin-exchange and excited state polarization transfer led to an $F = 4$ ground state polarization of approximately 70%.

To detect an EDM, the electric field is applied to the z axis, in which case an EDM would lead to a precession of the x polarization onto the y axis, as shown in (3.17). The analysis of the y polarization was accomplished by a 20 times weaker probe laser, tuned to the $6s_{1/2}, F = 4 \rightarrow 6p_{1/2}$ transition. The circular polarization of this beam was modulated at 33 kHz, by a photoelastic modulator, hence its average polarization was zero. A precession of the polarization onto the y axis would have been evident by a modulation in the intensity of the transmitted probe light in phase with, and at the frequency of, the modulation, and correlated with the application and sign of the electric field. A 1.2 μG field was applied along z to provide a calibration; the change in the y component of the polarization, in the presence of an EDM, is given by

$$\Delta P_y = 2p_x \omega_E = 4P_x d(\mathrm{Cs}) E\tau / [(2I+1)\hbar],$$

where ω_E is the EDM induced spin precession rate in the applied field E, τ is the spin polarization lifetime, and $I = 7/2$ is the nuclear spin. A dilution of the electron spin projection, because of the coupling to the nuclear spin, is evident by the factor $2I+1$ in the denominator. This reflects the fact that the electron spin changes by only $1/(2I+1)$ between m_F states, and thus the contribution of an energy associated with the electron spin direction is reduced by this factor.

The pseudoscalar that phenomenologically describes the sought experimental T violation is the precession angle of the atomic polarization about the electric field, and is proportional to

$$\boldsymbol{J} \cdot (\boldsymbol{E} \times \boldsymbol{\sigma})\tau,$$

where $\boldsymbol{J}$ is the angular momentum of the analyzing photons, $\boldsymbol{\sigma}$ the atomic polarization vector, and τ the spin relaxation lifetime. This quantity changes sign on reversal of $\boldsymbol{J}$, of $\boldsymbol{E}$, or of $\boldsymbol{\sigma}$. The factor $\boldsymbol{E} \times \boldsymbol{\sigma}$ describes the precession of the spin due to the EDM interaction; the factor τ gives the net precession angle. The net effect on the light is given by the projection of the precessed spin onto the light angular momentum, $\boldsymbol{J}$. An EDM would be evident by a change in transmitted light for fixed $\boldsymbol{J}$ and $\boldsymbol{\sigma}$ when $\boldsymbol{E}$ is reversed.

The most important systematics were those associated with imperfect reversal of the electric field coupling with the tensor polarizability of the Cs ground state, resulting from the hyperfine interaction of the $s_{1/2}$ ground state (Sect. 3.5.2). The small tensor polarizability of the Cs ground state leads to two scalar, T conserving quantities, $(\boldsymbol{\sigma} \cdot \boldsymbol{E})(\boldsymbol{J} \cdot \boldsymbol{E})$ and $(\boldsymbol{\sigma} \cdot \boldsymbol{J})E^2$. To the extent that $\boldsymbol{E}$ reverses precisely in magnitude, these terms do not mimic a true EDM. As is well known, perfect reversal of an electric field in the presence of dielectrics, contact potentials, etc. is difficult. The quality of the $\boldsymbol{E}$ reversal was determined by a precession about y from another polarizability term, $(\boldsymbol{\sigma} \cdot \boldsymbol{E})(\boldsymbol{\sigma} \times \boldsymbol{E})$, which does not reverse with the pump light helicity and could therefore be distinguished from other effects. By applying a 22 μG field along y, $\boldsymbol{\sigma} \cdot \boldsymbol{E}$ could be increased, enhancing these various terms, and limits could be set on their possible contributions to the EDM signal due to the imperfect reversal of $\boldsymbol{E}$. The application and reversal of B_y was directly incorporated into the automated data acquisition.

Other systematic effects include the possibility of a $\boldsymbol{v} \times \boldsymbol{E}$ field; this can come about because there is a spatial inhomogeneity of the atomic polarization, hence a directed diffusion of atoms in the cell. This effect, which would be evident as a magnetic field perpendicular to z (recall that there is no static applied magnetic field, so effects along each axis enter fully), was found to be unimportant, as were several other possible systematic effects.

6.5 The Tl EDM Experiment

After nearly 20 years of minimal interest, work with Tl was resumed by a group at Berkeley. Tl has a high enhancement factor of ≈ -600. The modern version of this experiment employs laser optical pumping and detection, instead of magnetic state selectors and hot-wire detectors; the various improvements have led to a six orders of magnitude increase in sensitivity over the 1970 experiment [219, 220].

A schematic of the experiment is shown in Fig. 6.1. In this experiment, two counter-propagating beams are used to discriminate the motional field effect from a possible true EDM. The two beams do not run simultaneously; every few minutes, the beam direction is changed (by opening and closing the appropriate shutters). Other experimental parameters are also varied; the electric field polarity is reversed about every second. The beams propagate vertically; this avoids possible complications associated with parabolic trajectories. Most systematic effects are relatively insensitive to the actual x velocity, while transverse velocity components are quite troublesome.

Natural Tl is used; there are two isotopes, 203 (30%) and 205 (70%). Both isotopes have nuclear spin-1/2, and the magnetic moments are very nearly equal; however, only the more abundant 205 isotope is used. The $F = 0$ and $F = 1$ ground state hyperfine splitting is 21 GHz. As shown in Fig. 6.2, the allowed optical electric dipole transition at 378 nm is used for optical state

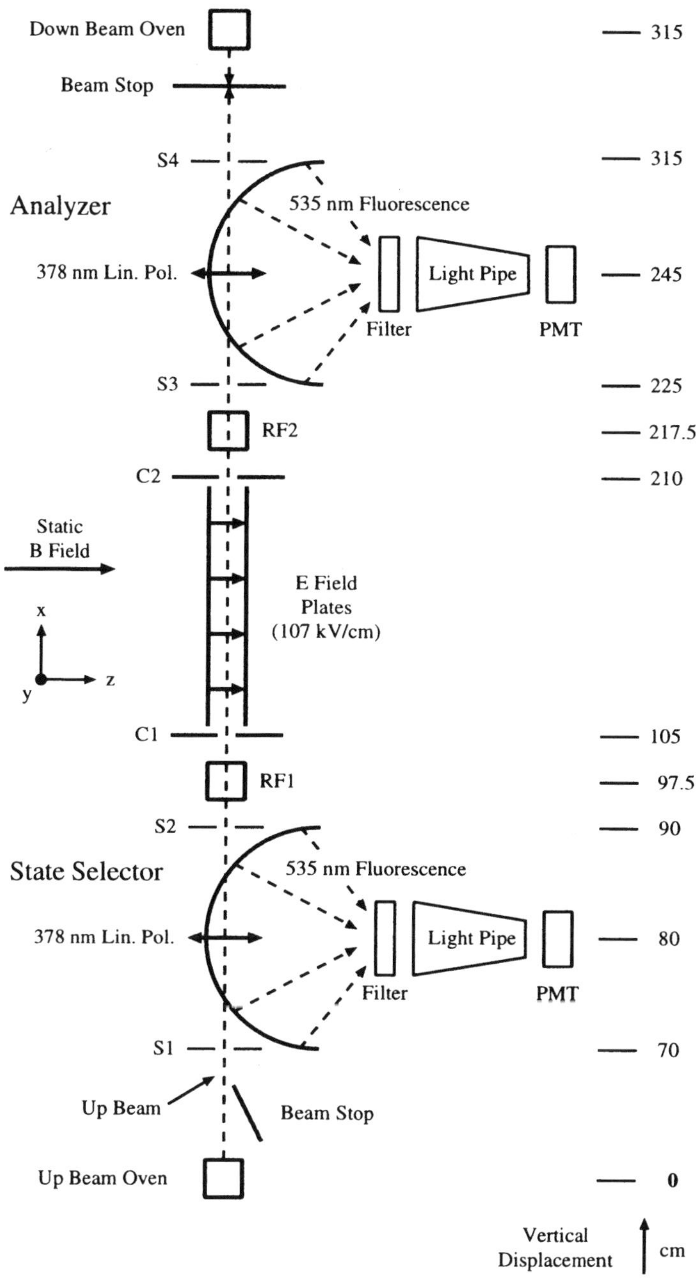

Fig. 6.1. Schematic of the Berkeley Tl EDM dual-beam experiment; C1,2 and S1–4 are collimators and slits. The laser beam is directed into the page, with linear polarization vector parallel to $\boldsymbol{B}$

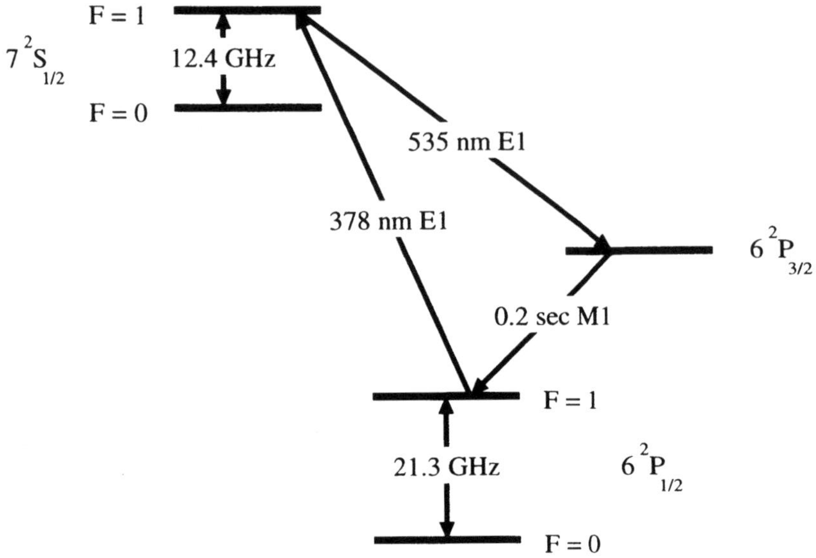

Fig. 6.2. Low-lying Tl atomic energy levels

selection; fluorescence from the decay of the $7s$ state is used for optically detecting whether transitions have occurred in the interaction region.

Consider the beam propagating upwards in Fig. 6.1. Tl atoms effuse from an electrically heated stainless steel oven, through a rectangular slit, 0.125 cm wide, 1 cm long, and 1 cm deep. The oven temperature is 1000 K, but slow atoms are scattered out of the beam, so the apparent temperature after the slit is 1100 K, corresponding to a most probably velocity of 4×10^4 cm/s, giving an interaction region coherence time of 2.5 ms. The flux was typically 10^{17} atoms/s.

The beam of $6p_{1/2}$ ground state atoms moves along the x axis, into a magnetically shielded region (four layers) in which a uniform magnetic field B_z of order 0.4 G or less is applied. The atoms first encounter a laser beam directed along y with linear polarization along z, tuned to the 378 nm $F = 1 \to F = 1$ transition, as shown in Fig. 6.3. Ground state atoms in the $|1,1\rangle$ and $|1,-1\rangle$ hyperfine sublevels make transitions to the $7s$ $|1,1\rangle$ and $|1,-1\rangle$ hyperfine states, respectively, while atoms in the $|1,0\rangle$ ground state have no absorption, because the Clebsch–Gordan coefficient for the corresponding transition vanishes; $\langle 1,0,1,0,|1,0\rangle = 0$. The excited state can decay to any three of the $F = 1$ ground state sublevels, or to the $F = 0$ ground state, or to the $6p_{3/2}$ level with emission of 535 nm fluorescence. (A large polished ellipsoidal mirror with focus at the atom–laser intersection collects a good fraction of the fluorescent light and directs it into a light pipe and to a photomultiplier for detection.) All of the atoms are pumped out of the $|1,1\rangle$

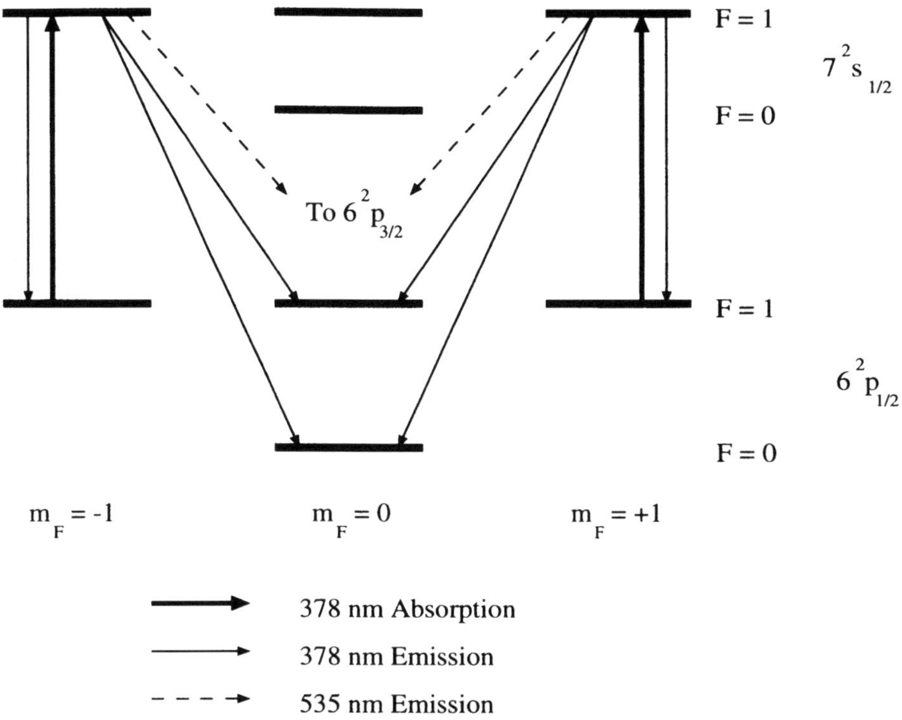

Fig. 6.3. Schematic of the optical pumping process. The polarization of the laser beam is selected to drive the $\Delta m = 0$ transitions

and $|1, -1\rangle$ hyperfine sublevels. All $F = 0$ states and the $6p_{3/2}$ (0.2 s lifetime) state can henceforth be ignored.

The net result is that a pure state $|1, 0\rangle$ is produced. These atoms then pass through the first coil of a separated oscillatory field RF spin flipper where a nearly resonant oscillating magnetic field in the x direction induces transitions. Instead of using a full quantum mechanical description of the system, some insight can be gained if we crudely model the $|1, 0\rangle$ state as two antiparallel spin-1/2 particles, as shown in Fig. 6.4 (see, for example, Appendix E of [51]). Then we can consider the two particles independently, and apply the formalism developed in Sect. 3.1.2 directly. After the $\pi/2$ pulse, the spins are perpendicular to the static field, and precess freely in magnetic and electric fields (of order 100 kV/cm) over a path length of 100 cm, after which a second RF flip coil is encountered. As before, if the second coil is in phase with the first, the transition probabilities are only quadratic in the detuning. Unlike the single spin case, a quadratic sensitivity also results if the second pulse is $\pi/2$ out of phase with the first. This can be seen by noting that when B_1 (the static field in the rotating frame) is parallel to the spin, the degree to which the $|1, 0\rangle$ state is repopulated must be quadratic in the accumulated small phase offset. A linear sensitivity results if an odd multiple of $\pi/4$ phase shift is introduced between the coils.

More formally, the initial state can be described by the following wave function (identical in the laboratory and rotating frames)

$$\Psi_1 = \begin{pmatrix} 0 \\ 1 \\ 0 \end{pmatrix}$$

and after the RF coil, in the rotating frame,

$$\Psi_2 = -\frac{\mathrm{i}}{\sqrt{2}} \begin{pmatrix} 1 \\ 0 \\ 1 \end{pmatrix}.$$

If there is an EDM (or mistuning of the RF frequency), the ± 1 state will pick up a phase shift,

$$\Psi_3 = -\frac{\mathrm{i}}{\sqrt{2}} \begin{pmatrix} \mathrm{e}^{-\mathrm{i}\epsilon_0} \\ 0 \\ \mathrm{e}^{\mathrm{i}\epsilon_0} \end{pmatrix} \approx -\frac{\mathrm{i}}{\sqrt{2}} \begin{pmatrix} 1 - \mathrm{i}\,\epsilon_0 \\ 0 \\ 1 + \mathrm{i}\,\epsilon_0 \end{pmatrix},$$

where, for an EDM-induced frequency offset,

$$\epsilon_0 = -\frac{d(\mathrm{Tl}) E_z}{\hbar} \frac{L}{v}.$$

After the second RF coil, including a phase offset of $(2n+1)\pi/4$, the wavefunction in the rotating frame is

$$\Psi_4 = \begin{pmatrix} \pm\frac{1}{2}\mathrm{e}^{\mp\mathrm{i}\,\pi/4}(1 \mp \epsilon_0) \\ \frac{1}{\sqrt{2}}(1 \pm \epsilon_0) \\ \pm\frac{1}{2}\mathrm{e}^{\pm\mathrm{i}\,\pi/4}(1 \mp \epsilon_0) \end{pmatrix},$$

where the upper sign refers to $B_z > 0$. The relative fraction of atoms in the $|1,0\rangle$ state does not change on transforming back to the laboratory frame.

The atoms then enter the analyzer, essentially identical to the state selector, and any atoms in the $|1,\pm 1\rangle$ states are excited. The 535 nm fluorescence signal is proportional to the sum of those populations:

$$S = |1 \mp \epsilon_0|^2 = 1 \mp 2\epsilon_0.$$

This signal is monitored as E_z is reversed, and the asymmetry

$$\Delta = \frac{S(+E_z) - S(-Ez)}{S(+E_z) + S(-E_z)}$$

is determined. This asymmetry is proportional to the P,T-odd rotational invariant $\boldsymbol{E}\cdot\boldsymbol{B}$. The experimental goal is to measure Δ and thereby determine $d(\mathrm{Tl})$ and thus d_{e}.

Figure 6.5 shows a plot of the fluorescence signal as the RF field frequency is varied. The rate at which the fringe visibility decreases with frequency

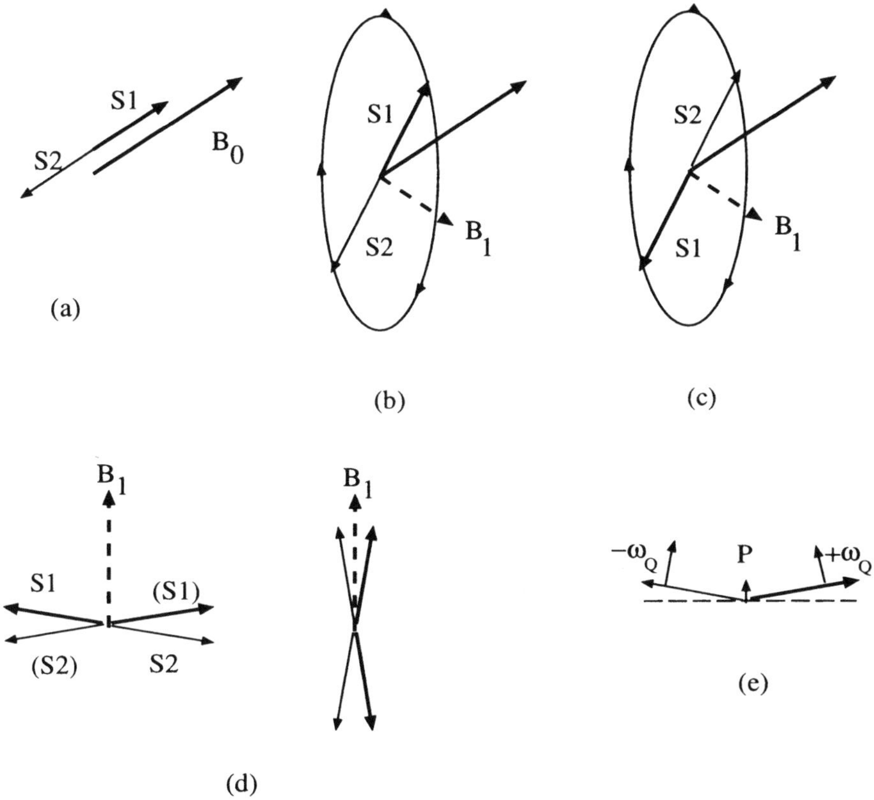

Fig. 6.4a–e. Crude model of the state used in the Tl beam experiment. (a) Two antiparallel spins, along B. (b) After an RF pulse, the spins precess freely. (c) After π rad (one-half cycle), the initial state is reached. This gives a factor of two reduction in net line width. (d) With a $\pi/2$ phase shift, a quadratic sensitivity results because small offsets in opposite directions are indistinguishable. (e) The experimental technique is relatively insensitive to tensor polarizability effects; such effects rotate the pure alignment into a vector polarization which is not discriminated by the linearly polarized detection light. Under operation of the experiment, the total precession due to ω_Q is on the order of 10 rad for atoms with the most probable velocity. The net effect is that the magnitude of the pure alignment is reduced by a factor of roughly $\langle|\cos\theta|\rangle = 2/\pi$ due to the averaging over the finite width velocity distribution of the beam; possible systematics associated with ω_Q are thereby substantially reduced. The loss in alignment is compensated by the effective factor of two reduction in linewidth and this technique therefore has an intrinsic efficiency of approximately $2 \times 2/\pi = 1.27$

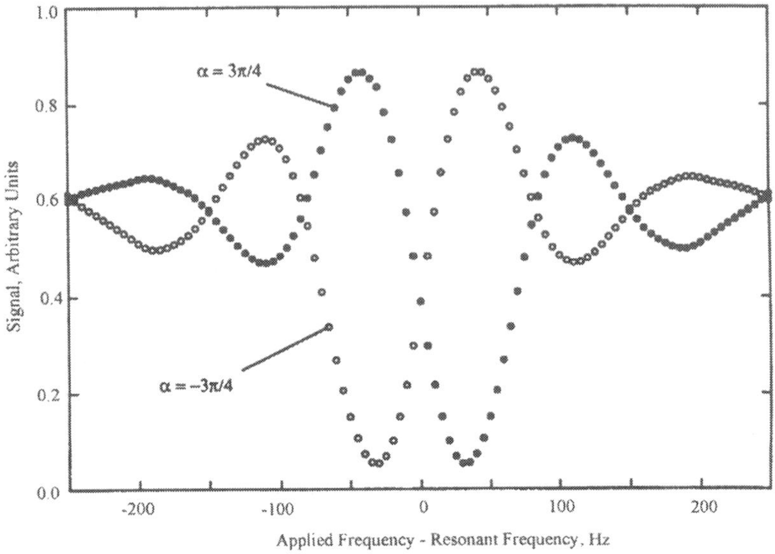

Fig. 6.5. Fluorescence signal from the Berkeley Tl EDM experiment

detuning is a measure of the beam velocity spread, as previously described [see (3.10)].

A number of noise sources limited the experimental sensitivity; these included shot noise on the photoelectrons, atomic beam intensity fluctuations, small spatial motion of the laser beams, etc. For the 1990 data, the analyzer signal at resonance was 2×10^7 photoelectrons/s, and was improved to 10^8/s for the 1994 result. However, the result was very close to the shot noise limit; using (3.5) with one day of data accumulation ($t = 86400$ s) and a coherence time $T = 2.4$ ms leads to (with an electron EDM enhancement of 600)

$$d_e < 6 \times 10^{-28}\, e\,\text{cm}.$$

The experiment was also limited by how well possible systematic effects could be eliminated, the most important of which were motional field effects, contributing 6×10^{-28} $e\,$cm (one sigma) to the electron EDM error. The second major source of possible systematic error was a geometric phase effect; this represents a coupling of the motional field to a gradient in the x direction. The idea is that the motional field (along y) turns on adiabatically as the atoms enter the electric field region; if there is a changing field along x (due to a small $\partial B_x/\partial x$ gradient), the tip of the net magnetic field vector traces out a closed path in space. This results in an additional phase shift of opposite sign for the $|\pm 1\rangle$ states. Possible inability to control this effect contributed 4.8×10^{-28} e cm (one sigma) to the electron EDM error. The final result was reported as

$$d_e = (1.8 \pm 1.2 \pm 1.0) \times 10^{-27}\, e\,\text{cm} \longrightarrow d_e < 4 \times 10^{-27}\, e\,\text{cm}, \tag{6.5}$$

where the first error is statistical, and the second reflects how well possible systematic effects could be eliminated.

6.6 Future Prospects for Improving the Electron EDM Limit

At present, the electron EDM limits from Cs are about a factor of 5 lower than the published 1989 result. However, with the improved accuracy, spurious and nonstatistical noise is contaminating the data ($\chi^2 > 1$ for data taken between major adjustment to the apparatus). In order to get a handle on polarizability and magnetic leakage effects, another alkali metal, perhaps Rb, could be contained in the same optical pumping cell. The density would have to be low enough that the spin-exchange frequency shifts between the two species are acceptably low; this can be easily achieved with present operating conditions (the spin exchange rate should be much less than the spin coherence time) [223].

The Tl beam experiment is currently being rebuilt, incorporating Na into the Tl oven, making a dual species beam. The Na will allow independent measures of the motional field and geometric phase effects [224]. Na has $F = 1$ and $F = 2$ hyperfine levels in the ground state, which have g values that are equal in magnitude but opposite in sign. The motional field effects are of opposite sign for these two states, while the geometric phase effect does not depend on g but only on m_F, so has the same sign between the two states and is twice as large for $F = 2$. By performing the experiment alternately on these two states, enough information is obtained to solve the two simultaneous equations with two unknowns, and in principle will allow a clean separation between the motional field and geometric phase effects. Given the low density of atoms in the beam, spin interactions between the two species should be absent to high accuracy. Factors of 10 increases in both light collection/detection efficiency and beam flux have also been incorporated. Hopefully, this experiment will push the electron EDM experimental limit well into the 10^{-28} e cm range.

Other experimental proposals have considered the use of optically trapped and cooled atoms. A particularly attractive idea is to use the "Zacharias fountain" (see [51], Sect. V.5.2), with clouds of ultra-cold atoms launched vertically between electric field plates. As proposed, the atoms are launched upward from a trap, pass through a separated oscillatory field coil, pass upward between electric field plates, then fall back through, again into the separated oscillatory field coil, and are then optically analyzed. A flight path of up to 1 m has been suggested, giving a coherence time of about 0.5 s. This gives a linewidth of 1 Hz, compared to 20 Hz or more in the present cell experiments, and 2.4 ms in the beam experiment. It is estimated that 10^9 Cs atoms could be collected in a trap, launched, and subsequently detected with 10% efficiency. In one day of averaging, assuming an electric field of 100 kV/cm, an atomic EDM of 10^{-26} e cm results, with a factor of 100 reduction due to the Cs enhancement for the electron EDM. This is 40 times better accuracy than the current limit. On the other hand, not even rough calculations or estimates of possible systematic effects pertaining to such an experiment have been published [225]. Many of the issues associated with the development of a frequency standard [226], e.g., collision induced frequency shifts, are not important for an EDM experiment.

However, a fountain EDM experiment will be subject to its own class of problems. One insidious effect is an electric field effect on the atom trajectories; for such slow atoms, the energy associated with the scalar polarizability is comparable to the kinetic energy of the atom [227]. Small electric gradients between the field plates, creating forces on the atoms that do not change exactly on field reversal, can lead to systematic effects in the presence of magnetic gradients. Also important are magnetic field gradients *transverse* to the beam. These will limit the linewidth, and the average integrated magnetic field must not differ by more than 0.1 μG for the various possible Cs atom trajectories. For a beam cross-section of 5 mm, this is quite a severe constraint.

Performing an EDM experiment directly within an optical trap has also been suggested. In this case, the polarizability problems (associated with the optical transitions) that were encountered in the storage cell alkali experiments, thankfully absent in the fountain technique, reappear with a vengeance. Also problematic are forces associated with inevitable field gradients which change the equilibrium position of the atoms.

Some new ideas for a beam experiment with metastable atoms, Sm in particular, have been put forward [228]. Estimation of systematic effects and a general experimental description exist; in principle, a factor of 1000 improvement in the electron EDM limit appears feasible. However, estimation of the enhancement due to close lying opposite parity states in Sm is difficult, and in fact the enhancement may not exist at all. Ra is a more promising candidate; the calculations are quite reliable in this case, but given the intense radioactivity of Ra, any possible experiment seems very difficult. Metastable states of molecular PbO are also a possibility, and here too the calculations appear to be reliable.

Also of interest is the storage of paramagnetic atoms in frozen rare gases. A review of recent work is presented in [229]. Cs contained in a solid ^{4}He lattice has shown a spin polarization lifetime of order 1 s. However, a spin precession has not yet been measured, and this system might have a very short spin coherence time. The problem is that a slight imperfection in the sphericity of the "bubble" that contains the Cs atom would lead to an enormous quadrupole field, and this could certainly have some random component between Cs atom sites. The tensor polarizability shift between the Zeeman ground state levels, even with a very slight asymmetry, could be enormous, considering that the crystal field results in a shift of the optical pumping transition by about 10 nm (4×10^{12} Hz) [230]. Although the shift in optical transition frequency reflects a scalar polarizability effect, the effective implied electric field is enormous. This electric field will certainly have gradients due to bubble asymmetries, and can therefore couple to the small Cs ground state tensor polarizability, giving enormous shifts between the ground state m_F levels. Such a large random quadrupole shift could completely dominate and randomize any precession about a homogeneous static magnetic field [77]. However, this is simply idle speculation, and a spin precession experiment should be performed.

6.7 EDM Limits of Some Other Elementary Particles

We believe that experimental upper limits on EDMs of other particles should also be presented, and this seems to be the most appropriate place to discuss them; some of these experiments employed techniques that are similar to those that have been used in regard to the electron EDM.

6.7.1 The Proton

As discussed in Chap. 8 and 9, the best limits for the proton EDM come from atomic experiments, and the current limit from the TlF beam experiments is $d_\text{p} < 10^{-22}$ e cm. The earliest limit, $d_\text{p} < 1.3 \times 10^{-13}$ e cm, was obtained from analysis of the hydrogen spectrum [4, 231]. This limit was improved in 1960 by an experimental search for an anomalous precession of 150 MeV protons in a magnetic field, with the proton spin direction determined by scattering from a carbon target [232]. A limit of 3×10^{-15} e cm was obtained.

6.7.2 The Neutrino

A neutrino EDM would lead to anomalous ionization in a detector through its interaction with atomic electrons. By analyzing the Cowan-Reines neutrino experiment of 1957 [233], an upper limit for the neutrino EDM (times a form factor F) of $d_\nu F < 2 \times 10^{-20}$ e cm was obtained [234].

6.7.3 The Muon

Several experiments have set limits on the EDM for both positive and negative muons. These experiments were done in conjunction with $g-2$ measurements for the muon at the CERN muon storage ring. One of these limits was obtained by attributing any possible deviation between the experimental value for $g-2$ and the quantum electrodynamic theoretical expectation to a muon EDM. This resulted in a limit of $d_\mu < 0.74 \times 10^{-18}$ e cm [235]. With a modified experimental procedure, a direct experimental limit of $d_\mu < 1.05 \times 10^{-18}$ e cm was obtained [236].

6.7.4 The Λ^0 Hyperon

The Λ^0 hyperon is produced with partial polarization which can be determined from the asymmetries in the decay $\Lambda^0 \to p\pi^-$. Even though there is no externally applied electric field, in the rest frame of the hyperon moving in a static magnetic field $\boldsymbol{B}$, there will be an effective electric field $\boldsymbol{v} \times \boldsymbol{B}$. If the hyperon has an EDM, it would interact with this electric field [237]. From such an analysis, a limit of $d_\Lambda < 8 \times 10^{-15}$ $e\,$cm was obtained. Subsequently, with an improved apparatus, a limit $d_\Lambda < 1.5 \times 10^{-16}$ $e\,$cm was obtained [238].

6.7.5 The τ Lepton

By analysis of the angular distribution of the τ pairs in the electron–positron annihilation, an upper limit for the EDM (times a form factor F) of $d_\tau F(q^2 = 1225\,\mathrm{GeV}^2) < 1.4 \times 10^{-16}$ e cm was obtained [239]. The data on the radiative decay $Z \to \tau^+\tau^-\gamma$ lead to the constraint $d_\tau < 6 \times 10^{-16}$ e cm [240]. Under specific assumptions on the properties of hypothetical CP-odd interactions of the τ lepton, a limit of 5×10^{-17} e cm was obtained [241].

7. EDM Experiments with Diamagnetic Atoms

What is time? A mystery, a figment – and all-powerful. It conditions the exterior world, it is motion married to and mingled with the existence of bodies in space, and with the motion of these. Would there then be time if there were no motion? No motion if no time? We fondly ask.

Thomas Mann

7.1 Shielding in the $^1S_0$ System

The enormous enhancement of the atomic EDM from an electron EDM only works with paramagnetic atoms. However, relativistic and finite size effects are also at work in heavy diamagnetic atoms, and the shielding effects discussed in the last chapter are not as severe as one might expect. Furthermore, diamagnetic atoms have a number of advantages over paramagnetic atoms. For a diamagnetic atom, the EDM effect is associated with the nuclear spin direction, and given that the nuclear magnetic moment is of order 1000 times smaller than the electron moment, diamagnetic atoms are much less susceptible to systematic effects associated with $\boldsymbol{v} \times \boldsymbol{E}$ fields, leakage currents, magnetic fields, and systematic magnetic fields associated with high voltage switching.

Diamagnetic atoms with nuclear spin $I = 1/2$ can have, in suitable optical pumping cells, spin coherence times of several hundred seconds or longer. This is because the spin-1/2 system cannot couple to the electric field gradients that the atom experiences on collision with the cell walls. On the other hand, alkali atom optical pumping spin coherence times are on the order of 20 ms, limited by diffusion and collisions with the cell walls, or spin disorienting collisions with a buffer gas. This time is even smaller for beam experiments. Thus, the effective spin coherence time for a diamagnetic atom is a factor of about 10^4 greater than that for a paramagnetic atom.

The atomic theory of diamagnetic atoms is fully discussed in Chap. 8; a brief overview of the results will be presented here. An electron EDM can induce a diamagnetic atom EDM by coupling through the hyperfine interaction. This effect scales as $Z^2\alpha^2$. In this case, there is an additional reduction factor of $\mu_\mathrm{n}/\mu_\mathrm{e}$, the ratio of the nucleon to electron magnetic moments. For ^{199}Hg, with an additional relativistic enhancement, the atomic EDM is only of order 0.01 of the electron EDM; the limit on d_e from that experiment is 5×10^{-26} e cm, about an order of magnitude worse than the Tl beam experiment, and comparable to the Amherst Cs experiment. For the Tl experiment, the electric field was ten times larger, and the effective counting rate was 10^8/s as opposed to 10^9/s for the ^{199}Hg experiment. These factors, taken together, give a relative sensitivity Q to an electron EDM of

$$Q = \frac{100\ \mathrm{kV/cm}}{10\ \mathrm{kV/cm}} \frac{0.003\ \mathrm{s}}{100\ \mathrm{s}} \frac{600}{0.01} \sqrt{\frac{10^8}{10^9}} \approx 6,$$

which is in good enough agreement with the observed factor of 10. Thus we see that indeed experiments with diamagnetic atoms, even with not so favorable relativistic enhancement of electronic T violation effects, can be competitive with paramagnetic atom experiments. The loss in enhancement is made up for by insensitivity to magnetic perturbations and an increased spin coherence time. Also absent are the tensor polarizability effects that complicate the alkali experiments.

The primary advantages of diamagnetic atoms pertain to nuclear T violation studies. Because the intrinsic line widths are much narrower, the sensitivity to such effects are much greater than for paramagnetic atoms. As was discussed already, finite size effects and magnetic interactions spoil the shielding effect expected for a purely electrostatically bound system of point particles. Magnetic interactions dominate for light atoms, and for hydrogen, the atomic EDM resulting from a nuclear (proton) EDM d_N is

$$d(\mathrm{A}) \sim \alpha^2 \frac{m_\mathrm{e}}{m_\mathrm{p}} d(\mathrm{N}), \tag{7.1}$$

where m_e and m_p are the electron and proton masses. This effect increases roughly as Z^2. For heavy atoms, the volume effect (Schiff moment) dominates, with a contribution of order

$$d(\mathrm{A}) \sim RZ^2(R_0/a)^2 d(\mathrm{N}), \tag{7.2}$$

where R_0 is the nuclear radius, a the Bohr radius, and R a relativistic enhancement factor of order 10 for $Z = 80$. The enhancement with increasing Z results because only the electronic wave function within the nucleus contributes to the effect.

Only three stable isotopes with $^1S_0$ electronic configuration, and $I = 1/2$, are suitable for room temperature optical pumping work; these are ^{3}He, ^{129}Xe, and ^{199}Hg. The sensitivity of ^{3}He to T violating effects is expected to be quite small; nonetheless, it finds application as a comagnetometer, and is therefore relevant to our discussion.

7.2 The ^{129}Xe EDM Experiment

Results from this experiment, performed in Seattle, were reported in 1982 [242]. The technique was based on spin-exchange pumping of ^{129}Xe with optically pumped Rb. This process had already been extensively studied [243–245]. The basic idea is that polarization is transferred from the optically pumped Rb electrons to the Xe nucleus during collisions; the free precession of the Xe can be monitored by its back-effect on the Rb polarization. Unfortunately, direct optical pumping of Xe requires a vacuum ultraviolet source.

The optical pumping cells for this experiment were fabricated from pyrex or fused silica tubes (2 cm inside diameter, 2 mm thick walls, 1 cm long) to which metal coated (platinum) glass disks were glued with vacuum epoxy. The tube served as the dielectric spacer. Several different wall coatings were used; the wax dotriacontane was effective. Such coatings were not critical in achieving a satisfactory ^{129}Xe lifetime.

After evacuating the cell, a small amount of Rb was distilled into it, then 2 torr of natural Xe was introduced, followed by 200 torr N_2 which was necessary for the spin-exchange pumping; the N_2 also served as a buffer gas for the Rb optical pumping and as a high voltage insulator.

The magnetic moments of ^{129}Xe and ^{85,87}Rb (the constituents of natural Rb) will interact if they are brought close enough together, with precession of the spins about a common axis: this tends to equalize their average polarizations. If the average angle of precession $\phi \ll 1$ during one interaction, then the mutual change in polarization during one interaction is proportional to ϕ^2. If the Rb is polarized through optical pumping, the polarization is transferred to the ^{129}Xe through this process.

This transfer of polarization can occur in binary collisions; however, Rb and Xe form Van der Waals molecules, and the spin exchange rate within the molecule is about 100 times the binary exhange rate [246]. Since energy and momentum cannot be conserved for a binary collision resulting in a bound state, a three-body collision is required to form a molecule; the N_2 buffer gas serves the role of the third collision partner. The molecule formed is very weakly bound and generally is destroyed in the next collision. It was found that the spin-exchange cross section was a maximum at 15 torr N_2, too low for application of electric fields [245].

The experiment required at least a few hundred torr of background gas for high voltage stability, and as a Rb spin diffusion buffer gas. However, it was necessary to limit the Xe density in order to prevent rapid depolarization of the Rb. Furthermore, the required Rb density was such that the cell diameter would give roughly one absorption length for the optical pumping light. With higher pressure N_2 (200 torr), the effectiveness of the spin exchange is reduced. This can be understood as follows. The rate of molecule formation, in the limit of high N_2 pressure, is proportional to the product of the densities of the three reaction partners:

$$\Gamma_{\mathrm{mol}} \propto \rho_{\mathrm{N}_2}\rho_{\mathrm{Xe}}\rho_{\mathrm{Rb}}.$$

However, because the molecules are destroyed by collisions with the N_2, the lifetime is proportional to $1/\rho_{\mathrm{N}_2}$, and ϕ is proportional to the molecular lifetime. Therefore, rate of polarization change Γ_{Xe} for the Xe atoms is then

$$\Gamma_{\mathrm{Xe}} = \Gamma_{\mathrm{mol}}\phi^2 \propto 1/\rho_{\mathrm{N}_2},$$

so the rate of change of the Xe polarization is lower at higher N_2 densities. For about 200 torr, with Rb density giving 2 cm absorption length for D_1 optical pumping light by the Rb (which aligns the Rb electron spin), $\Gamma_{\mathrm{Xe}} \approx 1/1000$ s. This time corresponds to the typical ^{129}Xe spin relaxation times, of order 1000 s. These values are representative of the usual experimental conditions.

The 129 isotope constitutes 26% of natural Xe. Molecule formation with the other Xe isotopes leads to a loss of Rb polarization. Also, the Rb spin couples to the molecular rotations, and relaxes the Rb polarization independently of the spin-exchange [245, 246]. The Xe density is chosen so that the net rate of Rb polarization relaxation roughly matches all other Rb relaxation rates. Under these conditions, about 1/100 of the Rb polarization is ultimately transferred to the ^{129}Xe nuclei.

When the ^{129}Xe and Rb are polarized, the angle ϕ has a well-defined nonzero average, and represents the precession of the spins about each other. If the ^{129}Xe nuclei are aligned with a static magnetic field, and if the Rb spins are precessing in this field, the precession frequency will be different for the two orientations of the ^{129}Xe spins relative to the fixed field. Thus, the ^{129}Xe atoms, when polarized, present a small magnetic field to the Rb atoms, of order 10 μG under typical experimental conditions. This corresponds to a Rb precession frequency of about 5 Hz and is easily observed.

The experiment was operated as follows. A stack of three identical cells were placed in a three-layer magnetic shield (shielding factor 40 000 for transverse fields), with their axes perpendicular to the shield axis. Circularly polarized D_1 light from a semiconductor diode laser illuminated the cells; initially, an approximately 1 mG magnetic field was applied along the shield axis (z axis). After approximately 1000 s, the ^{129}Xe nuclei reached their equilibrium polarization, and the longitudinal field was switched off. A smaller transverse field of 0.1 mG (x axis) parallel to the applied electric field, generated by a cosine distribution winding, remains on, and the ^{129}Xe spins precess in this field at a frequency of about 0.1 Hz.

The transverse field is limited to 0.1 mG so that the Rb could be operated as a "zero-field" magnetometer, as described in Sect. 3.2.1 [see (3.17)]. The field maximum was determined by the requirement that the Rb spin should not precess through an appreciable angle in a spin lifetime, which was about 0.3 ms; this corresponds to a field of order 0.1 mG.

The ^{129}Xe spin precession was monitored by the optically pumped Rb zero-field magnetometer. Since the ^{129}Xe was precessing in the y–z plane parallel to the incident light direction z (along which the Rb was principally polarized), the change in transmitted light intensity due to the precession of the Rb polarization about the oscillating y component of the ^{129}Xe polarization would be only quadratic in that angle; this is because the change in transmitted light intensity is proportional to the cosine of the angle between the Rb polarization vector and the light axis, which is small in this case ($2\pi \times 5$ Hz $\times 0.3$ ms $= 0.01$ rad). However, this small angle can be measured with linear sensitivity if a small oscillating (200 Hz) magnetic field is applied along y, with amplitude such that the Rb polarization oscillates between maximum excursions of about 1 rad (this field does not affect the ^{129}Xe precession in any significant way). In the absence of a magnetic field along y, the transmitted light intensity will be modulated at twice the oscillation frequency; if a small static field is added along y, the Rb polarization oscillations become asymmetric about z, and this appears as a modulation at the oscillation frequency. Phase detecting the first harmonic signal then gives a measure of the effective magnetic field along y.

Thus, the free precession of the ^{129}Xe atoms could be directly monitored through the y component of their magnetization, as measured by the Rb zero-field magnetometers. This signal was monitored for the three cells for about 1000 s (the spin coherence lifetime) and the average precession frequencies for each were determined. Between the polarizing/free precession measurements, the electric fields within the cells were reversed by changing the polarity of the applied voltages. Fields of up to 5 kV/cm were applied. The three-cell stack was used to monitor drifts in the static precession field and to test for leakage current effects. The configuration of voltages was altered occasionally between data runs. Also, the circular polarization of the pump light was reversed occasionally.

The polarized Rb also generated an effective field for the ^{129}Xe, corresponding to a precession frequency of about 10 mHz; this field was perpendicular to the precession field and, insofar as this field was perpendicular to the light axis, would only affect the ^{129}Xe precession frequency quadratically. The concern is that, due to the Stark shift of the optical frequency used in pumping the Rb, the pumping rate, hence the effective Rb magnetization, could change. Since this effect is quadratic in the electric field, an asymmetry in the electric field strength could lead to a systematic effect. However, none was observed. Nor were any effects due to leakage currents evident.

The final result was

$$d(^{129}\mathrm{Xe}) = (0.2 \pm 1.1) \times 10^{-26}\ e\,\mathrm{cm} \longrightarrow d(^{129}\mathrm{Xe}) < 2 \times 10^{-26}\ e\,\mathrm{cm}, \quad (7.3)$$

where the error is purely statistical. The achieved accuracy is consistent with the observed signal-to-noise ratio for the y magnetization signal on the detected laser light. There was no estimate of the expected shot noise sensitivity. Although there were an enormous number of ^{129}Xe atoms in the system, the precession was observed through a feeble effect on the Rb atoms; the shot noise was thus determined by the photon flux. The Xe density was determined by the Rb spin relaxation rate which in turn determined the photon flux (the optical pumping rate should be approximately equal to the total Rb relaxation rate). The reduction in effective ^{129}Xe density is approximately given by the polarization transfer efficiency of 1/100, as described above. If we assume a detected light power of 0.3 mW, it would appear that the statistical limit above is one to two orders of magnitude greater than that expected from shot noise.

The only improvement that has been considered for this experiment is to operate the Rb magnetometer in a fashion such that the static magnetic field, in addition to the Xe spin precession, can be directly measured. However, at the required level of magnetometer sensitivity, the polarizability effects that plague the Cs Amherst experiment become important.

7.3 The ^{199}Hg EDM Experiment

The ^{199}Hg experiment is by far the simplest among the EDM experiments described in this book; the limit obtained on the ^{199}Hg atomic EDM is however the best of any. In fact, this may be the most sensitive experiment ever performed on any system, given that the detectable energy scale in regard to the spin precession frequency was of order 4×10^{-23} eV.

It is this simplicity that allowed such a strict experimental limit to be obtained. (One should remember that “simple” does not imply “easy”.) The entire experiment relied on rather old-fashioned, yet well-understood and controllable, experimental techniques. No lasers are to be found here; this has been a point of criticism of this work! In fact, it would not have been possible to achieve the levels of stability, both long- and short-term, with the use of a laser. The discharge lamp performed remarkably well, requiring no substantial adjustment over months of operation, and at the frequency of interest, was stable at the shot noise level.

The use of ^{199}Hg for an EDM experiment was proposed in 1982. Work on this atom was begun in Seattle in 1984, with a preliminary result reported in 1987 [247, 248]. This version of the experiment used an oscillating magnetic field driven atomic spin oscillator, as described in Sect. 3.2.1 [see (3.18)]. A major disadvantage of this technique is that it required a projection of the pump light onto the static magnetic field (quantization axis), and therefore the atoms experience an effective magnetic field induced by the polarized light [53]. This introduced noise through low-frequency drift of the pump light intensity, and could also introduce systematics; the optical transition frequency can change through the scalar polarizabilities of both the ground and excited states. The light shift is directly proportional to the difference between the pump light frequency and transition frequency.

In 1988, major improvements were made to the experiment, and a schematic of the final version is shown in Fig. 7.1. The improvements included incorporation of a modulated pumping scheme as described in Sect. 3.2.1 [see (3.21)–(3.23)]. This effectively eliminated the light shift for the two following reasons: the circular polarization of the light was modulated, and the quantization axis was perpendicular to the light propagation direction. Other improvements included use of isotopically enriched ^{199}Hg, stabilization of the optical pumping cell properties, and increased light detection efficiency [80, 249, 250].

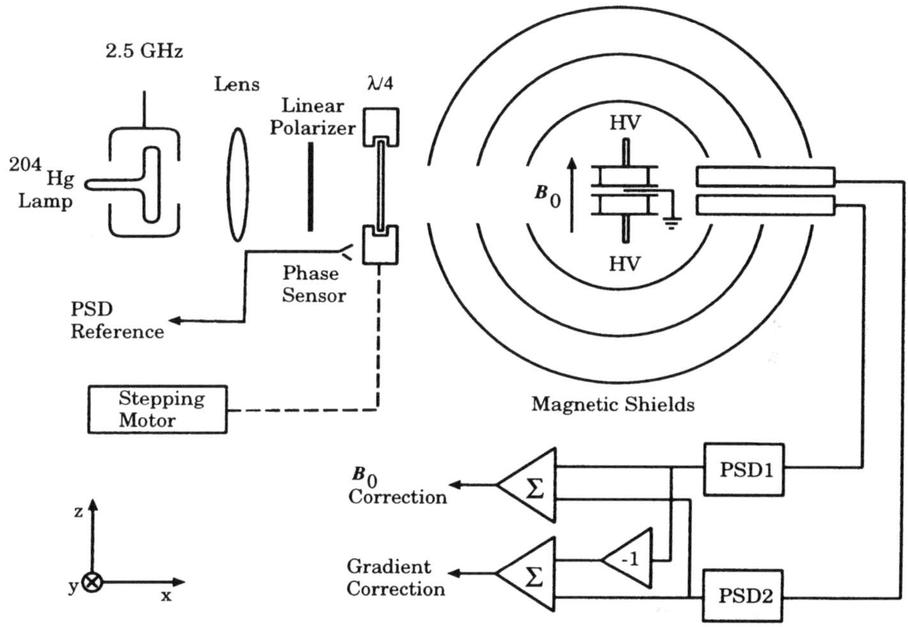

Fig. 7.1. Schematic of the ^{199}Hg EDM experiment

The optical pumping of Hg, and the transverse oscillator scheme is described in Sect. 3.2.1. As shown in Fig. 7.1, the circular polarization of the resonance light is modulated by a rotating quarter-wave plate. The modulator was a simple mechanical rotator driven by a stepping motor; a sixteen-slot phase pickup wheel was used as a reference for light intensity phase detectors, and there was no phase noise contributed from this system. This is because the experiment was operated in a two-cell differential mode; the phase noise, to very high accuracy, affects both atomic oscillators in the same way, and does not contribute to the differential signal.

The sum of the two oscillator out-of-phase light modulation component was used to stabilize the static magnetic field in a continuous feedback loop. The difference between the two signals was used to keep the magnetic gradient between the cells fixed; since the electric field is oppositely directed in the two cells, an EDM would be apparent as a change in the magnetic gradient correction signal.

Figure 7.2 shows the change in gradient as a function of time. The voltage applied to the cell electrodes is periodically reversed; the experimental goal was to detect a gradient change correlated with the electric field time pattern shown in this figure. The cells were contained in a three-layer magnetic shield (in fact the same shields as were used in the ^{129}Xe experiment described earlier) with transverse shielding factor of about 40 000. The transverse field was about 10 mG, corresponding to a Larmor frequency of 7 Hz. This frequency was chosen to be well away from the $1/f$ corner of the discharge lamp intensity noise spectrum.

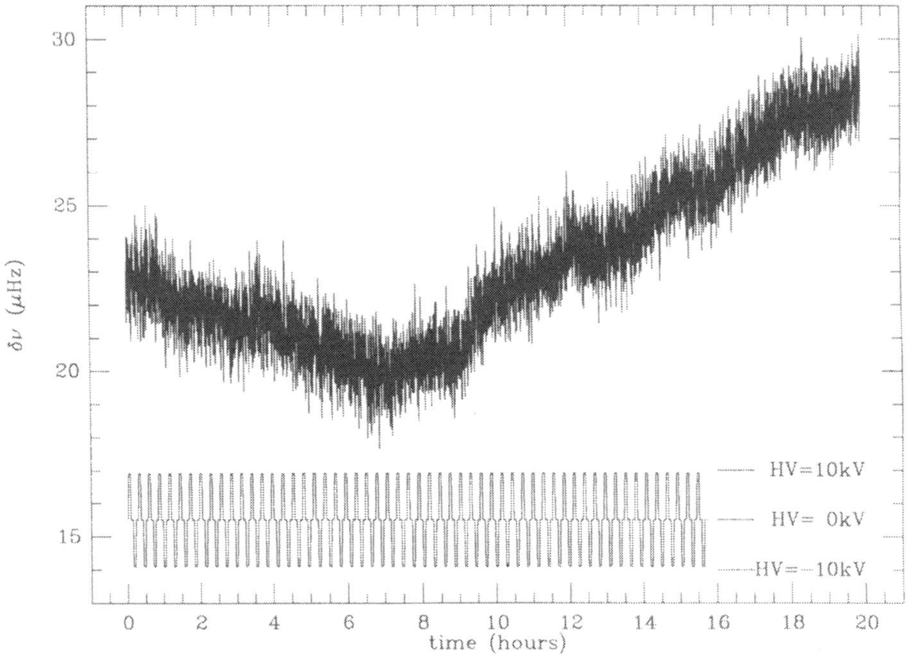

Fig. 7.2. Magnetic field gradient as a function of time. A ^{199}Hg EDM would be evident by a correlation of this signal with the applied electric field

The heart of the experiment was the optical pumping cells. These cells (shown schematically in Fig. 7.3) were in many ways similar to those used in the ^{129}Xe experiment and the Amherst Cs EDM experiment. However, there were some tricks required to get a stable, long ^{199}Hg lifetime. For the Cs experiment, the lifetime is limited by diffusion in the buffer gas, while the chemical inertness of ^{129}Xe means that it will not readily stick to the container walls in any case. However, Hg has some chemical activity and therefore tends to stick to the wall for a longer time, resulting in a more rapid spin dephasing. Therefore finding an appropriate wall coating to minimize the Hg binding energy, hence wall sticking collision dwell time, was a major problem. All previous work had employed heated (to over 300 °C) ultrapure fused silica [53]; an EDM experiment requires electrodes internal to the cell and should be operated at low temperature to minimize leakage currents which increase exponentially with temperature.

It was found that N_2 buffered cells with tin-oxide-coated glass electrodes, to which a thin coating of a paraffin wax was applied, and an ultrapure fused silica spacer, gave a reasonable lifetime; in fact, lifetimes up to 3000 s were observed. Unfortunately, in these original cells, the Hg would disappear. This was due to a slight O_2 impurity combining with the Hg in the presence of the ultraviolet pumping light. When the O_2 was removed, the cell lifetimes became unstable; the effect in this case was due to atomic hydrogen bound on the cell walls, as subsequent tests indicated. The hydrogen density was eventually controlled by introduction of 5% concentration of CO into the 200 torr N_2 buffer gas. Although the lifetimes were never as good as with pure N_2, the 70 s obtained was quite adequate, and in fact a longer lifetime would not have helped as the $1/f$ noise in the field gradient would have become important for a slow field reversal rate.

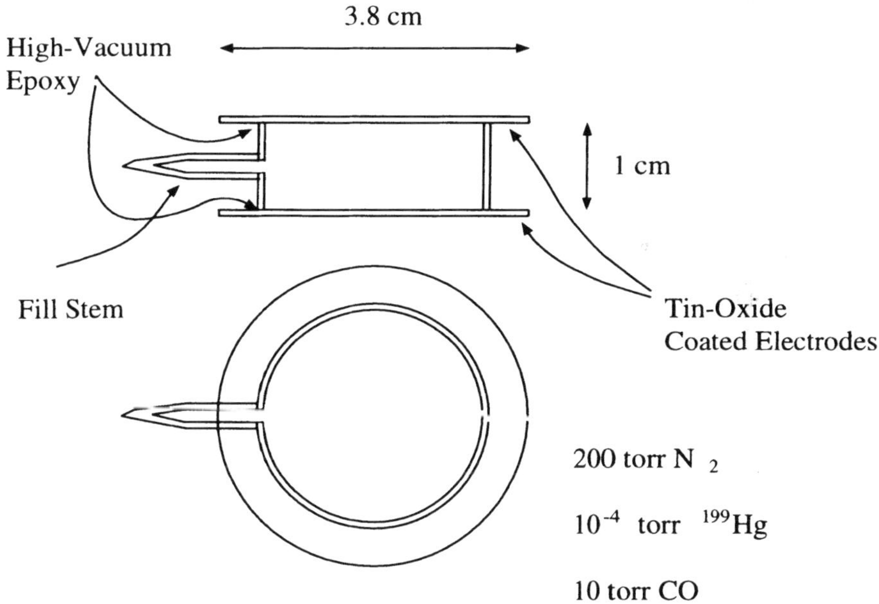

Fig. 7.3. Schematic of a cell used in the ^{199}Hg experiment

Only data taken in the last 1/5 of the period for a particular high voltage configuration was used in the EDM search. Since the atomic oscillators had some response time (about 20 s), it was necessary to wait for the system to respond to an EDM, and for possible magnetic effects associated with the field reversal charging currents to decay. There was no strong evidence for charging current effects. Also, leakage currents were at the 1 pA level, and would contribute a worst-case systematic about an order of magnitude below the statistical accuracy. The final result was reported as

$$d(^{199}\mathrm{Hg}) = (-1.0 \pm 2.6 \pm 3.5) \times 10^{-28}\ e\,\mathrm{cm}$$
$$\longrightarrow d(^{199}\mathrm{Hg}) < 8.7 \times 10^{-28}\ e\,\mathrm{cm}, \tag{7.4}$$

where the first error is statistical, and the second was the level at which a systematic error could be detected.

The statistical noise was at the shot noise level, and can be estimated as follows. The cell volume was 5 cm^3, and contained about 10^{-4} torr of ^{199}Hg, giving about 1.5 optical absorption lengths for the discharge lamp generated resonance light. This implies a total number of Hg atoms $N \approx 10^{13}$. As discussed in Sect. 3.2.1, the optimum sensitivity occurs when the pump rate is three times the relaxation rate (70 s), giving a photon flux onto the cell of

$$\Phi = 3N/T = 4 \times 10^{11}\ \mathrm{s}^{-1},$$

where T is the spin lifetime (the polarization lifetime and coherence time are equal under the experimental conditions). Of these photons, only about 1/5 were transmitted through the cell, and 1/4 of those were coupled to the photomultiplier which had quantum detection efficiency of 1/20; this gives an effective detected flux of $\Phi_\mathrm{d} = 1 \times 10^9\ \mathrm{s}^{-1}$. The atomic oscillator modulation amplitude was 12% of the transmitted light intensity; thus, the effective signal-to-noise was

$$\frac{\mathrm{S}}{\mathrm{N}} = 0.12 \times \sqrt{\Phi_\mathrm{d}} = 3800\,\sqrt{\mathrm{s}^{-1}}.$$

Taking a total lifetime of 20 s, which gives a line width of Δf =13 mHz, implies a frequency uncertainty of

$$\Delta f/(\mathrm{S/N}) = 4\ \mu\mathrm{Hz}\sqrt{\mathrm{s}}.$$

As shown in Fig. 7.2, where the data points represent a 10 s average of the frequency difference between the two cells (contributing a factor $\sqrt{2}$ to the observed noise), the RMS fluctuations are of order 2 μHz. Taking into account the 10 s average and the $\sqrt{2}$ factor, gives about 5 $\mu\mathrm{Hz}\sqrt{\mathrm{s}}$ noise, in agreement with the above estimate.

The EDM sensitivity was further reduced because 4/5 of the collected data, taken during the response waiting periods, was rejected. Also, zero-electric-field data, necessary for systematic error tests, did not contribute to the EDM signal. With these two additional loss factors, since 12% of the detected photons transmitted through the cell contribute to the signal, only about $1/10^4$ of the incident photons contribute to the EDM measurement.

Various schemes have been proposed to increase the sensitivity by about an order of magnitude, after which leakage current effects will become a problem. Also under consideration is an optically pumped comagnetometer, perhaps a spin-1/2 Cd isotope [247, 249]. Unfortunately, if Cd is used, the cell would have to be heated. Hg is too reactive to contain an alkali atom in the same cell; a ^{199}Hg – ^{129}Xe – Rb spin-exchange pumped scheme was found to be unworkable. In any case, in the light of the moderate systematic error in the present experimental limit, comagnetometer development can be deferred in anticipation of the next order of magnitude sensitivity increase.

7.4 ^{3}He – ^{129}Xe Comparison

Two species that can be simultaneously spin-exchange pumped are ^{3}He and ^{129}Xe. Since ^{3}He and Rb really do not form Van der Waals molecules, the spin-exchange transfer in this case is due to binary collisions, roughly 10 000 times slower than the ^{129}Xe rate [246, 251]. At 150 °C Rb equilibrium vapor density, the spin exchange time for ^{3}He is 22 h, compared to 7 s for ^{129}Xe, which increases to 12 min at 80 °C.

A first EDM experiment of this type was performed by applying electric field plates directly to a cell containing Rb, Xe, and ^{3}He [252]. The ^{3}He was first polarized over some hours with the cell maintained at 150 °C; since the T_1 relaxation time for ^{3}He is tens of hours, this polarization is done infrequently. At the high temperatures, the Rb polarization was reduced, and the ^{3}He polarization was limited to about 20%.

After the ^{3}He had acquired a good polarization, the temperature was reduced to 80 °C, giving a ^{129}Xe spin exchange rate comparable to the T_2 relaxation time. Then resonant oscillating field pulses were applied, $\pi/2$ rad for the ^{129}Xe, and about 0.1 rad for the ^{3}He. The idea was that just a little bit of the ^{3}He polarization would be used with each ^{129}Xe spin precession frequency measurement.

The freely precessing magnetization was measured by inductive pickup in broadly resonant coils. Since the signal amplitude is proportional to the oscillation frequency, the Larmor frequencies were set to about 3 kHz and 10 kHz for the ^{129}Xe and ^{3}He, respectively. The signal-to-noise ratio was of order 1000 $\sqrt{\mathrm{s}^{-1}}$, leading to a frequency uncertainty of about 0.1 μHz for a 1000 s free induction decay measurement. With the applied field of 2 kV/cm, this corresponded to an EDM of about 2×10^{-25} $e\,$cm.

Unfortunately, the high density strongly polarized ^{3}He presented an additional magnetic field to the ^{129}Xe. Because this field does not scale between the two species in the ratio of the magnetic moments (as for an external applied magnetic field), it represents a possible noise source. In fact, the decay of the ^{3}He polarization between successive free decay measurements limited the experimental accuracy (obtained in a long-term average) to a level a factor of 10 worse than expected from the observed noise. It should be emphasized that the anomalous shifts are not necessarily electric field dependent, but primarily introduce extra noise into the system.

In order to counter this effect, a steady state experiment based on Zeeman maser oscillation [253] was proposed and is presently being developed [254]. The idea is that the polarization and measurement are performed in separated cells, connected by a diffusion tube. The polarization cell is maintained at a high temperature, while the EDM measurement cell, where the maser oscillation takes place, is kept at or near room temperature. Leakage currents are thereby substantially reduced, and there is no Rb to relax the precessing spins. This geometry is slightly unfavorable in that the species with the fast diffusion time (^{3}He) has the long spin lifetime; nonetheless, a substantial fraction of the slowly diffusing ^{129}Xe polarization survives diffusion between the cells. In addition, the magnetic pickup coils must now be resonant with the precessing atoms. The pickup coils must be carefully made and temperature controlled because the maser frequency depends on the coil resonant frequency to some degree, leading to relative shifts that do not scale as true magnetic field changes. Also, variations in the atomic polarization can lead to magnetization induced drift as before. A primary goal is to decrease the amplitude, and increase the time scale, of these drifts. This technique has not yet proven successful; however, development is continuing as the preliminary results appear promising.

An interesting feature for the maser operation is that the frequency uncertainty decreases with averaging time t as

$$\Delta f \approx \frac{\sqrt{3}}{\pi} \Phi_{\mathrm{rms}} t^{-3/2}, \tag{7.5}$$

where Φ_{rms} is the (white) noise to signal amplitude ratio, per $\sqrt{\mathrm{Hz}}$ bandwidth. This is the same result that is obtained for the uncertainty in the determination of the oscillation frequency of a steady sinusoidal signal in the presence of noise [255]. An important point is that the atomic coherence time τ does not directly appear in this equation, but is implicit in Φ_{rms}. In the limit that Johnson noise V_{n} on the pickup coil dominates the system noise, Φ scales as $V_{\mathrm{n}}/\tau\Gamma_{\mathrm{p}}$, where τ is the relevant lifetime, and Γ_{p} is the polarization rate within the maser cell.

Because of the uncontrolled relative frequency shifts, the proposed maser experiment will require subtraction between repetitive reversals of the applied electric field; the rate at which this should be done is dictated by the rate of drift, but in a long time T, no more than T/t measurements can be done. This leads to $\Delta f \propto 1/t\tau$. In comparison to our previous analysis of frequency determination precision, Sect. 3.1.1, we can replace Φ_{rms} with $\sqrt{N/\tau}$ if we assume that all N particles of the system are dephased during the course of measurement; this gives the $1/\sqrt{\tau}$ behavior in the frequency accuracy obtained previously for a storage type experiment. It is apparent that for the maser, the extra factor $\sqrt{\tau}$ is due to the fact that the coherence is not destroyed during the course of a measurement over a time $t > \tau$.

8. Atomic Calculations

> A number, unknown to anybody,
> increased twice, looked at itself
> into a mirror and saw there 811.
> What had been this number before
> increasing twice?
>
> *Grigory Oster*

To obtain from the atomic experiments, described in the previous chapters, quantitative information on the dipole moments of elementary particles and on their CP-odd interactions, detailed atomic calculations must be performed. We confine ourselves in the present book to a simple, phenomenological approach to this problem. Its results are compared with those of more sophisticated atomic many-body calculations. Almost always the agreement between the two approaches is quite satisfactory; at least it is sufficient for the interpretation of the present experimental data. In some cases, when phenomenological calculations are lacking, we will resort to estimates and to results obtained within a more refined technique. Readers interested in the many-body approach to the calculation of atomic CP-odd effects are referred to the review [256].

8.1 Wave Function of an Outer Electron at Short Distances

The CP-odd interactions we are interested in are concentrated at short distances from the nucleus. In heavy atoms the motion of even a valence electron near the nucleus is relativistic. The relativistic wave function of an electron with principal quantum number n, total angular momentum j, orbital angular momentum l and projection of the total angular momentum m is (see, e.g., [31], §35):

$$u_{njlm} = \begin{pmatrix} f_{njl}\,\Omega_{jlm} \\ g_{njl}\,\mathrm{i}\,(-\boldsymbol{\sigma}\cdot\boldsymbol{r}/r)\,\Omega_{jlm} \end{pmatrix}. \tag{8.1}$$

Here Ω_{jlm} is a spherical function with spin, i.e., the eigenfunction of the operators $\boldsymbol{j}^2$ and j_z, constructed from the eigenfunctions of the operators $\boldsymbol{l}^2$

and l_z, which are ordinary spherical functions, and the eigenfunctions of s^2 and s_z, which are ordinary spinors. In particular, at $j_z = 1/2$ the two spin spherical functions are

$$\Omega_{1/2,0} = \frac{1}{\sqrt{4\pi}}\,\chi, \quad \Omega_{1/2,1} = \frac{1}{\sqrt{4\pi}}\,(-\boldsymbol{\sigma}\cdot\boldsymbol{r}/r)\,\chi. \tag{8.2}$$

As for the radial wave functions f_{njl} and g_{njl}, at distances $r \ll a/Z^{1/3}$, where the Coulomb field of the nucleus can be considered unscreened, while the electron binding energy can be neglected compared with the potential, they are expressed in terms of the Bessel functions as follows:

$$\begin{aligned} f_{njl}(r) &= \frac{c_{njl}}{r}\left[(\gamma + \kappa)\,J_{2\gamma}(x) - \frac{x}{2}\,J_{2\gamma-1}(x)\right], \\ g_{njl}(r) &= \frac{c_{njl}}{r}\,Z\alpha\,J_{2\gamma}(x). \end{aligned} \tag{8.3}$$

The following notation is used:

$$x = \left(\frac{8Zr}{a}\right)^{1/2}, \quad \gamma = \sqrt{(j+1/2)^2 - Z^2\alpha^2}, \quad \kappa = (l-j)(2j+1).$$

Expressions (8.3) can be verified by direct substitution in the Dirac system of radial equations (see [31], §35) at zero binding energy:

$$\begin{aligned} (rf)' + \frac{\kappa}{r}\,(rf) - \left(2m + \frac{Z\alpha}{r}\right)(rg) &= 0, \\ (rg)' - \frac{\kappa}{r}\,(rg) + \frac{Z\alpha}{r}\,(rf) &= 0. \end{aligned}$$

To determine the normalization constant c_{njl}, let us note that for $r \gg a/Z$ the radial function $f_{njl}(r)$ should go over into the ordinary semiclassical solution of the nonrelativistic Schrödinger equation

$$R_{nl}(r) = \frac{b_{nl}}{r\sqrt{k(r)}}\,\sin\left(\int_{r_1}^{r} dr'k(r') + \phi\right). \tag{8.4}$$

Here

$$k^2(r) = 2m\left[E_{nl} - U(r) - \frac{(l+1/2)^2}{2mr^2}\right],$$

r_1 is the turning point, and ϕ is a constant phase. The constant b_{nl} can be readily found (see, for instance, [257]). Indeed, the main contribution to the normalization integral is given by the semiclassical region lying between the two turning points r_1 and r_2, so that

$$b_{nl}^2 \int_{r_1}^{r_2} \frac{dr}{k(r)}\,\sin^2\left(\int_{r_1}^{r} dr'k(r') + \phi\right) \approx \frac{b_{nl}^2}{2}\int_{r_1}^{r_2}\frac{dr}{k(r)} = 1.$$

Differentiating the Bohr quantization rule for radial motion

$$\int_{r_1}^{r_2} \mathrm{d}r\, k(r) = \pi(n_\mathrm{r} + \beta)$$

with respect to n_r and taking into account that $k(r_{1,2}) = 0$, we get

$$m\,\frac{\mathrm{d}E_{nl}}{\mathrm{d}n_\mathrm{r}} \int_{r_1}^{r_2} \frac{\mathrm{d}r}{k(r)} = \pi .$$

Thus

$$b_{nl}^2 = \frac{2m}{\pi}\frac{\mathrm{d}E_{nl}}{\mathrm{d}n_\mathrm{r}} = \frac{2}{\pi a^2 \nu^3}.$$

We use here the phenomenological expression for the outer electron energy

$$E_{nl} = -\frac{m\alpha^2}{2\nu^2},$$

where ν is the effective principal quantum number: $\nu = n_\mathrm{r} + l + 1 - \sigma_l = n - \sigma_l$; the quantum defect σ_l is weakly n_r dependent at fixed l.

Comparing (8.3) with (8.4) at $a/Z \ll r \ll a/Z^{1/3}$, we obtain

$$c_{nlj} = \frac{\kappa}{|\kappa|}\left(\frac{\pi a}{2Z}\right)^{1/2} b_{nl} = \frac{\kappa}{|\kappa|}\left(\frac{1}{Za\nu^3}\right)^{1/2}. \tag{8.5}$$

Now the relativistic radial wave functions (8.3) can be written at short distances as follows:

$$f_{njl}(r) = \frac{\kappa}{|\kappa|}(\kappa - \gamma)\left(\frac{Z}{a^3\nu^3}\right)^{1/2} \frac{2}{\Gamma(2\gamma+1)}\left(\frac{a}{2Zr}\right)^{1-\gamma},$$

$$g_{njl}(r) = \frac{\kappa}{|\kappa|}Z\alpha\left(\frac{Z}{a^3\nu^3}\right)^{1/2} \frac{2}{\Gamma(2\gamma+1)}\left(\frac{a}{2Zr}\right)^{1-\gamma}. \tag{8.6}$$

In these expressions, the factor

$$\frac{2}{\Gamma(2\gamma+1)}\left(\frac{a}{2Zr}\right)^{1-\gamma}$$

is singled out because it goes to unity in the nonrelativistic limit (at $Z\alpha \to 0$) for $j = 1/2$. Let us note that, taking all the normalization constants b_{nl} to be positive, we have therefore made all the relativistic radial wave functions $f_{njl}(r)$ and nonrelativistic ones $R_{nl}(r)$ positive at $r \to 0$.

The following relations for the nonrelativistic radial wave functions of the s and p states at the origin are sometimes useful for concrete calculations:

$$R_{n0}|_{r\to 0} = 2\left(\frac{Z}{a^3\nu^3}\right)^{1/2}, \quad R_{n1}|_{r\to 0} = \frac{2}{3}\left(\frac{Z}{a^3\nu^3}\right)^{1/2}\frac{Z}{a}. \tag{8.7}$$

8.2 The Electron EDM in Paramagnetic Heavy Atoms

We now go over to a detailed calculation of the EDM of a paramagnetic heavy atom, as induced by the electron dipole moment. The covariant relativistic Hamiltonian (2.5) for the electron EDM interaction with electric field $\boldsymbol{E}$ can be written as a noncovariant perturbation

$$V_d = -d_e \gamma_0 \boldsymbol{\Sigma E}, \tag{8.8}$$

$$\boldsymbol{\Sigma} = \gamma_0 \gamma_5 \boldsymbol{\gamma} = \begin{pmatrix} \boldsymbol{\sigma} & 0 \\ 0 & \boldsymbol{\sigma} \end{pmatrix}.$$

It is convenient to rewrite this operator as

$$V_d = V_{1d} + V_{2d} = -d_e \boldsymbol{\Sigma E} - d_e(\gamma_0 - 1)\, \boldsymbol{\Sigma E}. \tag{8.9}$$

In a heavy atom the magnetic interaction between the electrons (spin–spin, spin–other-orbit, orbit–orbit), as well as the retardation effects in the Coulomb interaction between them, are small compared with the spin-orbit interaction which rapidly increases with Z (see, e.g., [258], §19). Neglecting the enumerated small effects, the Hamiltonian of a many-electron heavy atom can be written in the following relativistic form:

$$H_0 = \sum_k \left(-\mathrm{i}\boldsymbol{\alpha}_k \boldsymbol{\nabla}_k + m\,\gamma_{0,k} - \frac{Z\alpha}{r_k} \right) + \sum_{k>l} \frac{\alpha}{|\boldsymbol{r}_k - \boldsymbol{r}_l|}. \tag{8.10}$$

This Hamiltonian describes with good accuracy the structure of the levels of a heavy atom, including spin–orbit splitting. The idea of the representation (8.9) is that, as can be easily checked by direct calculation, in the absence of an external field the operator V_{1d} for each electron satisfies the identity

$$V_{1d} = \mathrm{i}\frac{d_e}{e}\, [\boldsymbol{\Sigma p},\ H_0].$$

But in this case, following the line of reasoning of Sect. 6.1, one can easily prove that the perturbation V_{1d} makes no contribution to the EDM of an atom at all. The remaining interaction V_{2d}, as it must, vanishes in the nonrelativistic approximation.

First consider the "bare" contribution of V_{2d} to the EDM of an atom:

$$d_e \langle n\, |(\gamma_0 - 1)|\, n \rangle. \tag{8.11}$$

One can easily see that only the external electrons, whose angular momenta are uncompensated, actually play a role in (8.11). However, for external electrons the matrix element in (8.11), arising only due to relativistic effects at distances $r \sim a$, has a negligible order of magnitude, $\alpha^2 \sim 10^{-4}$. Thus, the EDM of an atom is actually determined by mixing of the states of opposite parity due to the interaction V_{2d}. Simple calculations show it to be equal to

$$2\,d_{\mathrm{e}} \sum_m \frac{\left\langle n \left| \sum_l \boldsymbol{r}_l \right| m \right\rangle \left\langle m \left| \sum_k (\gamma_{0,k} - 1) \boldsymbol{\Sigma}_k \boldsymbol{\nabla}_k U(\boldsymbol{r}) \right| n \right\rangle}{E_n - E_m}, \tag{8.12}$$

where $U(\boldsymbol{r})$ is the total potential energy.

We will begin calculating the induced dipole moments of heavy atoms with the simplest case of cesium. A cesium atom has one valence electron outside the closed shells, which have the same structure as that of the preceding element – xenon. These shells are very compact, and therefore the intermediate states $|m\rangle$ in formula (8.12) can be reduced to excitations of the outer electron only, its ground state being $6s_{1/2}$. Then, in cesium, as is known from experiment (and reproduced by theoretical calculations), the amplitudes of the E1 transitions $6s - np$ are small for $n > 6$. Thus, the problem reduces to a single-electron one with the nearest excitation $6p_{1/2}$ of the valence $6s_{1/2}$ electron as the only intermediate state.

Now, the main contribution to the mixing matrix element comes from short distances, where the potential energy is due to the unscreened Coulomb field of the nucleus, $U(\boldsymbol{r}) = Z\alpha/r$, and for the radial wave functions one can use the approximation (8.3). A direct calculation results in the following expression for this matrix element for arbitrary total angular momentum j:

$$\begin{aligned} &\langle\, j,\, l = j \pm 1/2 \,|\, (\gamma_0 - 1) |\, Z\alpha(\boldsymbol{\Sigma}\boldsymbol{r})/r^3 \,|\, j,\, l = j \mp 1/2 \,\rangle \\ &\quad = -\frac{4(Z\alpha)^3}{\gamma(4\gamma^2 - 1)a^2(\nu'\nu)^{3/2}}. \end{aligned} \tag{8.13}$$

Due to the dependence of $\gamma = \sqrt{(j+1/2)^2 - Z^2\alpha^2}$ on Z, the matrix element (8.13), and with it the induced EDM of an atom, increases with Z somewhat faster than Z^3.

As to cesium ($Z = 55$), using the experimental values of the required energy interval (see the tables [259]),

$$E(6p_{1/2}) - E(6s) = 11178\,\mathrm{cm}^{-1},$$

the effective principal quantum numbers (extracted also from [259]),

$$\nu(6s) = 1.87,\ \nu(6p_{1/2}) = 2.33,$$

and the radial integral [260],

$$\rho(6s, 6p_{1/2}) = \frac{1}{a} \int_0^\infty \mathrm{d}r\, r^3 R_{60}(r) R_{61} \;=\; -\,5.535(14),$$

we find the enhancement factor for the electron EDM in this atom:

$$\begin{aligned} K(\mathrm{Cs}) &= \frac{d(\mathrm{Cs})}{d_{\mathrm{e}}} \\ &= -\frac{16}{3} \frac{Z^3\alpha^2\rho(6s, 6p_{1/2})}{\gamma(4\gamma^2-1)(\nu_{6p_{1/2}}\nu_{6s})^{3/2}} \frac{\mathrm{Ry}}{E(6p_{1/2}) - E(6s)} = 132. \end{aligned} \tag{8.14}$$

This result is in quite reasonable agreement with more refined calculations [207,209,211,261–265]. The most detailed among these calculations results in [211]

$$K(\mathrm{Cs}) = 114, \tag{8.15}$$

with an estimated error of around 3%.

So, the experimental result (6.4) for the dipole moment of the cesium atom gives for the electron EDM

$$d_e < 10^{-25}\ e\,\mathrm{cm}. \tag{8.16}$$

The state of art of modern atomic many-body theory allows one to reach an accuracy of around 1% in the theoretical value of $K(\mathrm{Cs})$. The same accuracy can be reached in atomic calculations of other CP-odd effects in cesium as discussed below.

The calculation of the enhancement factor for the electron dipole moment in thallium is more tedious. The thallium ground state is $6s^2\,6p_{1/2}$ (the superscript indicates, as usual, the number of equivalent electrons; two $6s$ electrons in this case), and the optical transitions here correspond to the excitation of not only $6p_{1/2}$, but also of $6s$ electrons. Another complication (for the phenomenological approach) is that the radial integrals of the E1 transitions $6p_{1/2} - ns$ decrease slowly with growing n. It is necessary therefore to allow for the admixture of a large number of the ns states to $6p_{1/2}$. The corresponding phenomenological calculation [208] resulted in

$$K(\mathrm{Tl}) = -500. \tag{8.17}$$

Its accuracy can be estimated as 10–15%.

The value of the enhancement factor in thallium obtained in [266] is considerably larger than (8.17), which can, in all probability, be explained by the overestimated contribution of the excitations of the $6s$ electron. The Hartree-Fock calculation in [263] resulted in $K(\mathrm{Tl}) = -1906$. Allowing for the first-order corrections, the same group obtained $K(\mathrm{Tl}) = -1041$ [264]. Further corrections calculated in [211, 267] reduced this value, below the phenomenological value (8.17). But the most extensive Hartree–Fock calculations [212] give a result

$$K(\mathrm{Tl}) = -585, \tag{8.18}$$

with the estimated accuracy 5–10%. There is obviously no contradiction between the phenomenological value (8.17) of the enhancement factor and this one.

It is instructive to discuss why the signs of K(Cs) and K(Tl) are different. The matrix element (8.13) of the effective EDM interaction is real, and therefore has the same sign for both the $s_{1/2} - p_{1/2}$ and the $p_{1/2} - s_{1/2}$ admixtures. All the energy intervals, $6s - np$ in cesium and $6p - ns$ in thallium, are also of the same sign. What does change is the sign of the dominant radial integrals. In Cs $\rho(6s, 6p)$ is negative, which can be easily explained. According to our convention, both radial wave functions are positive at the origin. But their radial quantum numbers n_r, together with the number of nodes, differ by unity. So, at large distances the radial functions have opposite signs. No wonder therefore that the radial integral, controlled just by large distances, is negative. Analogous arguments demonstrate that $\rho(6p, 7s)$, dominating in thallium, is positive. And for another excitation of importance in Tl, $6s - 6p$, the negative sign of its contribution to K(Tl) is due to the Pauli principle (see, e.g., [23], pp. 68,97).

Now, with the enhancement factor (8.18), the experimental result for the dipole moment of the thallium atom, obtained in [220], leads to the best upper limit for the electron EDM (6.5)

$$d_e < 4 \times 10^{-27}\, e\,\text{cm}. \tag{8.19}$$

For the $^3P_2$ state of the xenon atom the EDM enhancement factor is [208, 210]

$$K(\text{Xe}^*) = 120, \tag{8.20}$$

so that the experimental upper limit obtained in [210] (see Table 6.2) gives

$$|d_e| < 3 \times 10^{-24}\, e\,\text{cm}. \tag{8.21}$$

8.3 CP-Odd Electron–Nucleon Interaction

It is expedient now to digress from the electron dipole moment for the discussion of atomic EDMs as induced by the CP-odd electron–nucleon interaction. If we restrict to four-fermion operators without derivatives [see (2.12)], this interaction can be written

$$H = \frac{G}{\sqrt{2}}\,(k_1 \bar{N}N\, \bar{e}\,\mathrm{i}\gamma_5 e + k_2\, \frac{1}{2}\epsilon_{\kappa\lambda\mu\nu}\, \bar{N}\sigma_{\kappa\lambda}N\, \bar{e}\sigma_{\mu\nu}e + k_3 \bar{N}\,\mathrm{i}\gamma_5 N\, \bar{e}e). \tag{8.22}$$

Here the dimensionless constants k_1, k_2 and k_3 characterize the strength of the hypothetical interaction. In the limit of an infinitely large mass of the nucleon, only two terms in (8.22) survive. They reduce, in the coordinate representation, to the noncovariant Hamiltonian

$$H = \mathrm{i}\,\frac{G}{\sqrt{2}}\,\delta(\boldsymbol{r})\,(k_1\gamma_0\gamma_5 + 2k_2\boldsymbol{\sigma}_N\boldsymbol{\gamma}). \tag{8.23}$$

The interaction of an electron with a nucleus of charge Z and neutron number N can now be expressed as

$$H_1 + H_2 = \mathrm{i}\,\frac{G}{\sqrt{2}}\,\delta(\boldsymbol{r})\left[(Zk_{1p} + Nk_{1n})\,\gamma_0\gamma_5 \right.$$
$$\left. + 2\left(k_{2p}\sum_p \boldsymbol{\sigma}_p + k_{2n}\sum_n \boldsymbol{\sigma}_n\right)\boldsymbol{\gamma}\right]. \tag{8.24}$$

8.3.1 *CP*-Odd Mixing of Atomic Levels

Mixing of atomic levels due to interaction (8.24) can be calculated by means of the wave functions (8.3). The only nonvanishing matrix element refers to the mixing of the $s_{1/2}$ and $p_{1/2}$ states. The divergence that arises in the calculation of the matrix element of $\delta(\boldsymbol{r})$ means in fact that the nucleus cannot be regarded as pointlike. The finite size of the nucleus is introduced as a cut-off of the divergent integrals:

$$\int \mathrm{d}\boldsymbol{r}\, \delta(\boldsymbol{r})\, r^{-\epsilon} = R_0^{-\epsilon}, \quad \epsilon > 0.$$

For the nuclear radius R_0 we assume the usual dependence on atomic number A:

$$R_0 = r_0 A^{1/3}, \quad r_0 = 1.2\ \mathrm{fm}.$$

Now the mixing matrix element is

$$\begin{aligned}\langle s_{1/2}|H|p_{1/2}\rangle = \frac{Gm^2\alpha^2}{\sqrt{2}\,\pi}\frac{Z^2R}{(\nu_s\nu_p)^{3/2}}\,\mathrm{Ry}\Big\{&\gamma\,(Zk_{1p}+Nk_{1n})\\ &-4\boldsymbol{j}\,\frac{2+\gamma}{3}\left\langle k_{2p}\sum_p \boldsymbol{\sigma}_p + k_{2n}\sum_n \boldsymbol{\sigma}_n\right\rangle \\ &+4\boldsymbol{j}\,(1-\gamma)\left\langle k_{2p}\sum_p\left[\boldsymbol{n}_p(\boldsymbol{\sigma}_p\cdot\boldsymbol{n}_p)-\frac{1}{3}\boldsymbol{\sigma}_p\right]\right.\\ &\left.+\,k_{2n}\sum_n\left[\boldsymbol{n}_n(\boldsymbol{\sigma}_n\cdot\boldsymbol{n}_n)-\frac{1}{3}\boldsymbol{\sigma}_n\right]\right\rangle\Big\}.\end{aligned} \tag{8.25}$$

Here R is the relativistic enhancement factor defined so that $R \to 1$ as $Z\alpha \to 0$:

$$R = \frac{4}{\Gamma^2(2\gamma+1)}\left(\frac{a}{2ZR_0}\right)^{2-2\gamma}. \tag{8.26}$$

This factor increases rapidly for large Z, ranging from 2.7 and 2.8 for xenon and cesium ($Z = 54, 55$) to 8.2 and 8.5 for mercury and thallium ($Z = 80, 81$). The unit radius vectors of the protons and neutrons, $\boldsymbol{n}_p$ and $\boldsymbol{n}_n$, arise in (8.25) from the electron coordinate $\boldsymbol{r}$, since in this local interaction the electron and nucleon radii must coincide.

The brackets $\langle\ldots\rangle$ in the matrix element (8.25) denote averaging of the corresponding structures over the nuclear state with given angular momentum $\boldsymbol{I}$. Clearly, both vectors with which the angular momentum $\boldsymbol{j}$ of the electron is contracted are directed along $\boldsymbol{I}$. The first nuclear matrix element in (8.25)

$$\left\langle k_{2p}\sum_p \boldsymbol{\sigma}_p + k_{2n}\sum_n \boldsymbol{\sigma}_n\right\rangle$$

will usually be estimated in the simple nuclear shell model: a valence nucleon above a closed shell. The error of this approximation can reach 30%.

The term in (8.25) containing

$$\left\langle k_{2p} \sum_p \left[\boldsymbol{n}_p(\boldsymbol{\sigma}_p \boldsymbol{n}_p) - \frac{1}{3}\boldsymbol{\sigma}_p \right] + k_{2n} \sum_n \left[\boldsymbol{n}_n(\boldsymbol{\sigma}_n \boldsymbol{n}_n) - \frac{1}{3}\boldsymbol{\sigma}_n \right] \right\rangle,$$

enters with a relatively small factor $\sim Z^2\alpha^2$, and usually is omitted. Moreover, in some cases, when high accuracy cannot be claimed anyway, we will put $\gamma = 1$ in other terms of (8.25) as well.

It is instructive to consider the mixing of atomic states caused by this interaction. Let for instance a $p_{1/2}$ state be admixed to a $s_{1/2}$ one. In the resulting wave function

$$\frac{1}{\sqrt{4\pi}}[R_0(r) - \eta R_1(r)(\boldsymbol{\sigma} \cdot \boldsymbol{n})]\chi, \tag{8.27}$$

constructed with the spherical functions with spin (8.2), the mixing coefficient η is real, together with the matrix element (8.25). Obviously, in the state with the wave function (8.27) the charge density of the electron is shifted with respect to the plane $z = 0$ either upwards or downwards, depending on the spin projection on the z axis. In other words, due to the T-odd interaction the atom has acquired a dipole moment. On the other hand, in this state the spin acquires a component directed along the radius-vector $\boldsymbol{r}$, even when the electron is unpolarized in the absence of mixing. As a result, we get a peculiar spin hedgehog [268] (see also [269]). Curiously, the corresponding distribution of the magnetic moment creates no magnetic field.

But let us come back to atomic experiments discussed in the previous chapters. Their results can be used to obtain upper limits on the constants k_1 and k_2 of the CP-odd electron–nucleon interaction (8.22) [270–277].

8.3.2 Paramagnetic Atoms

We will begin with cesium. Elementary calculations show that the state arising from the ground state as a result of the discussed mixing, is

$$|\widetilde{6s_{1/2}, F}\rangle = |6s_{1/2}, F\rangle - 3.7 \times 10^{-11} \Big\{ (0.41 k_{1p} + 0.59 k_{1n})$$

$$+ 0.74 \times 10^{-2} \left[F(F+1) - \frac{33}{2} \right] k_{2p} \Big\} |6p_{1/2}, F\rangle. \tag{8.28}$$

It is sufficient here to limit ourselves to the admixture of the $|6p_{1/2}, F\rangle$ state since, as has been noted in the previous section, for $n > 6$ the E1 amplitudes $6s - np$ are small. The spin of the ^{133}Cs nucleus is $I = 7/2$, so that with an electron angular momentum $j = 1/2$ the possible values of the total angular momentum of the atom are $F = 3,\ 4$. To calculate the contribution to the mixing from the tensor interaction with the constant k_2, we use the shell model of the nucleus, according to which in ^{133}Cs the state of the unpaired proton is $g_{7/2}$. In the atomic state (8.28) the EDM is

$$\begin{aligned} d(\mathrm{Cs}) &= \langle \widetilde{6s_{1/2}, F} | - ez | \widetilde{6s_{1/2}, F} \rangle \\ &= 1.34 \times 10^{-10} \Big\{ (0.41 k_{1p} + 0.59 k_{1n}) \\ &\quad + 0.37 \times 10^{-2} \left[F(F+1) - \frac{33}{2} \right] k_{2p} \Big\} ea. \end{aligned} \tag{8.29}$$

The result for the scalar contribution (dependent on $0.41k_{1p} + 0.59k_{1n}$), obtained in [271], coincides numerically with the corresponding term in (8.29), though the analytical formula used in [271] for the mixing matrix element, differs from (8.25) in having the factor γ^2 instead of γ in front of $0.41k_{1p} + 0.59k_{1n}$. A similar result was obtained also by a many-body technique in [277].

Neglecting the contribution of the tensor interaction, we obtain from the experimental result (6.4) for the dipole moment of the cesium atom the upper limit

$$0.41 k_{1p} + 0.59 k_{1n} < 1.6 \times 10^{-5}. \tag{8.30}$$

On the other hand, if we assume that the scalar interaction is absent, i.e., if we set $0.41k_{1p} + 0.59k_{1n} = 0$, and take account of the fact that the constraint (6.4) is obtained for the atomic state with $F = 4$, formula (8.29) yields the following result for the tensor constant:

$$k_{2p} = 1.3 \times 10^{-3}. \tag{8.31}$$

Let us pass over now to thallium. The calculation using formula (8.25), which differs but slightly from those for the electron EDM enhancement, gives the following value for the dipole moment of this atom as induced by interaction (8.22):

$$d(\mathrm{Tl}) = -ea \times 0.96 \times 10^{-9} \left[(0.4 k_{1p} + 0.6 k_{1n}) - 10^{-3} k_{2p} \right]. \tag{8.32}$$

A refined many-body calculation [256, 277] predicts a larger numerical factor in this formula: 1.30 ± 0.04. With this larger factor the experimental result (6.5) leads to the limit

$$0.4 k_{1p} + 0.6 k_{1n} < 3.4 \times 10^{-7}. \tag{8.33}$$

We obtain here also

$$k_{2p} < 3.4 \times 10^{-4}. \tag{8.34}$$

However, the relative coefficient of k_{2p} in (8.32) is not only anomalously small [compare with (8.29)], also its value is not sufficiently reliable. The reason for both facts is the strong cancellation in the tensor interaction between the contributions of the excitations of the $6p$ and $6s$ electrons.

8.3.3 Diamagnetic Atoms

Much better upper limits on the constant k_2 can be obtained from the results with diamagnetic atoms, xenon and mercury. As we see from expression (8.25) for the matrix element, the interaction independent of the nuclear spin, i.e., that determined by the constants $k_{1p,n}$, does not open closed electron shells and therefore by itself induces no EDM in the ground state of Xe and Hg. Let us consider the contribution of the nuclear-spin-dependent CP-odd interaction, starting with xenon. In the shell model an unpaired neutron in the ^{129}Xe nucleus is in the $s_{1/2}$ state. In this approximation the first nuclear matrix element in formula (8.25), $\langle k_{2p} \sum_{\mathrm{p}} \boldsymbol{\sigma}_p + k_{2n} \sum_n \boldsymbol{\sigma}_n \rangle$, reduces to $k_{2n}\boldsymbol{I}/I$, while the second one vanishes. We will perform the atomic part of the estimate following [278]. The excited states of the xenon atom lie high up, and the ionization potential is close to one Rydberg. In this case it is natural to expect that the largest matrix elements of the radius r will correspond to transitions to the states of the continuous spectrum with $E \sim$ Ry. Therefore, the principal effective quantum numbers will be $\nu_s \sim \nu_p \sim 1$. Let us further assume the typical value of the radial integral to be $\rho \sim 1$. (Note that the estimate of xenon polarizability using these numbers gives reasonable agreement with experiment.) Under those assumptions, the estimate for the EDM of a diamagnetic atom, as induced by the tensor interaction, is

$$d \sim \frac{Gm^2\alpha^2 Z^2 R}{\sqrt{2}\,\pi}\, 4ea \left| \left\langle k_{2p} \sum_{\mathrm{p}} \boldsymbol{\sigma}_p + k_{2n} \sum_n \boldsymbol{\sigma}_n \right\rangle \right| . \tag{8.35}$$

To obtain the numerical factor 4 in this estimate, we put $2|\boldsymbol{j}| = 1$ in formula (8.25) for the mixing, and take into account that this mixing takes place in both wave functions entering the expectation value. For ^{129}Xe ($Z = 54,\ R = 2.7$) formula (8.35) gives

$$d(^{129}\mathrm{Xe}) \sim 0.6 \times 10^{-20} \left| \left\langle k_{2p} \sum_p \boldsymbol{\sigma}_p + k_{2n} \sum_n \boldsymbol{\sigma}_n \right\rangle \right| e\,\mathrm{cm}. \tag{8.36}$$

In the simple shell model the state of the valence neutron in the ^{129}Xe nucleus is $s_{1/2}$, so that the nuclear matrix element in formula (8.36) reduces to k_{2n}. This value can be refined [274, 279] if we neglect the contribution of the spin–orbit interaction to the configuration mixing in the ^{129}Xe nucleus. In such an approximation the ground state of this nucleus, despite the configuration mixing, has the "naive" quantum numbers: $L = 0,\ S = 1/2$. Then, from the magnetic moment of the nucleus

$$\mu(^{129}\mathrm{Xe}) = -0.77 = -1.91 \left\langle \sum_n \sigma_{n,z} \right\rangle + 2.79 \left\langle \sum_p \sigma_{p,z} \right\rangle ,$$

and the obvious relation

$$\left\langle \sum_n \sigma_{n,z} + \sum_p \sigma_{p,z} \right\rangle = 1,$$

we obtain

$$\left\langle k_{2p} \sum_p \sigma_{p,z} + k_{2n} \sum_n \sigma_{n,z} \right\rangle = (0.24 k_{2p} + 0.76 k_{2n}) \frac{\boldsymbol{I}}{I}. \tag{8.37}$$

Detailed Hartree–Fock calculations for the xenon atoms [273, 274] give results which are in a reasonable agreement with each other and with the naive estimate (8.36):

$$d(^{129}\mathrm{Xe}) = 0.52 \times 10^{-20}\, k_2(^{129}\mathrm{Xe})\, e\,\mathrm{cm}, \tag{8.38}$$

and

$$d(^{129}\mathrm{Xe}) = 0.41 \times 10^{-20}\, k_2(^{129}\mathrm{Xe})\, e\,\mathrm{cm}, \tag{8.39}$$

respectively. Taken together with the experimental constraint (7.3), these numbers result in

$$k_2(^{129}\mathrm{Xe}) = 0.24 k_{2p} + 0.76 k_{2n} < 4 \times 10^{-6}. \tag{8.40}$$

Under the same assumptions for the Hg atom as for Xe, the estimate (8.35) gives for mercury

$$d(^{199}\mathrm{Hg}) \approx 1.3 \times 10^{-20}\, k_{2n}\, e\,\mathrm{cm}. \tag{8.41}$$

We have calculated here the nuclear matrix element in the simple shell model, where the ^{199}Hg nucleus contains a neutron in the $p_{1/2}$ state above the closed shells, so that

$$\langle \boldsymbol{\sigma}_n \rangle = -\frac{1}{3} \frac{\boldsymbol{I}}{I}.$$

The result of Hartree–Fock calculations for mercury [273] is (in the same nuclear shell model)

$$d(^{199}\mathrm{Hg}) = 2.0 \times 10^{-20}\, k_{2n}\, e\,\mathrm{cm}. \tag{8.42}$$

Now the experimental bound (7.4) gives

$$k_{2n} < 4 \times 10^{-8}. \tag{8.43}$$

From the experiments with ^{129}Xe and ^{199}Hg one can also derive quite strict limits on the constant k_3 in the interaction (8.22). In the lowest nonvanishing approximation in m_p^{-1}, the noncovariant Hamiltonian of the corresponding electron–nucleus interaction reduces to the form

$$H_3 = -\frac{G}{\sqrt{2}} \frac{1}{2m_\mathrm{p}} \left(k_{3p} \sum_p \boldsymbol{\sigma}_p + k_{3n} \sum_n \boldsymbol{\sigma}_n \right) \boldsymbol{\nabla} \delta(\boldsymbol{r})\, \gamma_0. \tag{8.44}$$

The singularity of this operator in the limit of a pointlike nucleus is strong, so that the calculation of the matrix element of mixing of the $s_{1/2}$ and $p_{1/2}$

states requires here accurate treatment of the finite size of the nucleus. When estimating this matrix element, we will limit ourselves, however, to a simple model in which the CP-odd interaction is concentrated on the surface of a sphere of radius R_0. In this case the mixing matrix element is as follows:

$$\langle s_{1/2}|H_3|p_{1/2}\rangle = \frac{Gm^2\alpha^2}{\sqrt{2}\,\pi}\,\frac{Z^2R}{(\nu_s\nu_p)^{3/2}}\,\mathrm{Ry}$$

$$\times\left(-\frac{Z\alpha}{3m_\mathrm{p}R_0}\,2\boldsymbol{j}\left\langle k_{3p}\sum_p \boldsymbol{\sigma}_p + k_{3n}\sum_n \boldsymbol{\sigma}_n\right\rangle\right). \tag{8.45}$$

Comparing this expression with (8.25), we find the correspondence:

$$k_3 \leftrightarrow \frac{6m_\mathrm{p}R_0}{Z\alpha}k_2 = 4.6\times10^3\frac{A^{1/3}}{Z}\,k_2. \tag{8.46}$$

Thus, the experiments with ^{129}Xe and ^{199}Hg yield, respectively,

$$0.24k_{3p} + 0.76k_{3n} < 1.7\times10^{-3}, \tag{8.47}$$

and

$$k_{3n} < 1.4\times10^{-5}. \tag{8.48}$$

Moreover, from the experiments with the diamagnetic atoms we can obtain limits even on the constants $k_{1p,n}$ [242, 275, 276]. In particular, the sensitivity of the mercury experiment is so high that the thus obtained limit approaches that of (8.33). To switch on the CP-odd scalar interaction H_1 (the first term in Hamiltonian (8.24) which depends on the constants $k_{1p,n}$) in a diamagnetic atom, we have to allow for the hyperfine interaction which polarizes the closed electronic shells along the nuclear spin:

$$V_\mathrm{hf} = e\boldsymbol{M}\cdot\frac{\boldsymbol{r}\times\boldsymbol{\alpha}}{r^3}. \tag{8.49}$$

Here

$$\boldsymbol{M} = \frac{e\mu}{2m_\mathrm{p}}\,\frac{\boldsymbol{I}}{I}$$

is the magnetic moment of the nucleus, μ is its value in units of the nuclear magneton $e/2m_\mathrm{p}$, and $\boldsymbol{I}$ is the nuclear spin. The thus induced EDM of the atom arises in third order of perturbation theory and has the form

$$\sum_{n,k}\frac{\langle 0|-e\boldsymbol{r}|n\rangle\langle n|V_\mathrm{hf}|k\rangle\langle k|H_1|0\rangle}{(E_0-E_n)(E_0-E_k)} \tag{8.50}$$

$$+\ \text{permutations of operators.}$$

Let us note first of all that the corrections to the wave function of the initial state $|0\rangle$ generated by both operators V_hf and H_1 are concentrated at short distances from the nucleus. Therefore, among the permutations in equation (8.50) one can neglect those where $-e\boldsymbol{r}$ is placed in the middle, between these operators. When treating other terms in this expression, it is convenient to introduce the effective operator

$$\widetilde{H}_1 = \sum_k \frac{V_{\rm hf}|k\rangle\langle k|H_1 \; + \; H_1|k\rangle\langle k|V_{\rm hf}}{E_0 - E_k} \; . \tag{8.51}$$

Rather tedious calculations lead to the following expression for the only matrix element of this effective operator which is singular in the nuclear radius R_0:

$$\begin{aligned}\langle s_{1/2}|\widetilde{H}_1|p_{1/2}\rangle &= \frac{Gm^2\alpha^2}{\sqrt{2}\pi}\,\frac{Z^2R}{(\nu_s\nu_p)^{3/2}}\,\mathrm{Ry} \\ &\times \left[-\frac{2}{3}\,\frac{\alpha\mu}{m_{\rm p}R_0}\,(Zk_{1p} + Nk_{1n})\,2\boldsymbol{j}\cdot\boldsymbol{I}/I\right] .\end{aligned} \tag{8.52}$$

This result has been somewhat refined [276] by using a more realistic description of the magnetic moment density in the nucleus, i.e., of the hyperfine interaction at short distances, and of the density of nucleons in the nucleus. According to [276], the right-hand side of formula (8.52) should be supplemented by the factor $1.2(1 + 0.3Z^2\alpha^2)$. Comparing now (8.52) with (8.24), we obtain the correspondence

$$\begin{aligned}\left(k_{1p}\frac{Z}{A} + k_{1n}\frac{N}{A}\right)\frac{\boldsymbol{I}}{I} &\leftrightarrow \frac{3m_{\rm p}R_0}{A\alpha\mu}\,0.81(1+0.3Z^2\alpha^2)^{-1} \\ &\quad\times\left\langle k_{2p}\sum_p \boldsymbol{\sigma}_p + k_{2n}\sum_n \boldsymbol{\sigma}_n\right\rangle \\ &= 1.9\times 10^3\,(1+0.3Z^2\alpha^2)^{-1}A^{-2/3}\mu^{-1} \\ &\quad\times\left\langle k_{2p}\sum_p \boldsymbol{\sigma}_p + k_{2n}\sum_n \boldsymbol{\sigma}_n\right\rangle .\end{aligned} \tag{8.53}$$

Taken together with (8.40) and (8.43), as well as with the values of the magnetic moments of the nuclei [$\mu(^{129}\mathrm{Xe}) = -0.77, \;\; \mu(^{199}\mathrm{Hg}) = 0.50$], the last relation yields

$$0.41k_{1p} + 0.59k_{1n} \; < \; 4\times 10^{-4} \tag{8.54}$$

and

$$0.40k_{1p} + 0.60k_{1n} \; < \; 1.4\times 10^{-6} \tag{8.55}$$

for xenon and mercury, respectively.

8.3.4 Summary of the Constants $k_{1,2,3}$

The upper limits on the constants $k_{1,2,3}$ discussed in this section are summarized in Table 8.1. For completeness we also include here the results of the molecular TlF experiment discussed in the next chapter.

Table 8.1. Upper limits on the constants of CP-odd electron–nucleon interactions

Atom or molecule	k_1	k_2	k_3
Cs	1.6×10^{-5}	1.3×10^{-3}	
Tl	3.4×10^{-7}	3.4×10^{-4}	
^{129}Xe	4×10^{-4}	4×10^{-6}	1.7×10^{-3}
^{199}Hg	1.4×10^{-6}	4×10^{-8}	1.4×10^{-5}
TlF	1.7×10^{-5}	4.5×10^{-7}	3×10^{-4}

These results for the CP-odd electron–nucleon interaction can be transformed into the upper limits on the corresponding constants of the electron–quark interaction defined in close analogy to (8.22) as follows:

$$H_q = \frac{G}{\sqrt{2}} \left(k_{1q} \bar{q} q \, \bar{e} \, \mathrm{i} \gamma_5 e + k_{2q} \frac{1}{2} \epsilon_{\kappa\lambda\mu\nu} \, \bar{q} \sigma_{\kappa\lambda} q \, \bar{e} \sigma_{\mu\nu} e \right. \\ \left. + k_{3q} \bar{q} \, \mathrm{i} \gamma_5 q \, \bar{e} e \right). \tag{8.56}$$

As mentioned already in Sect. 5.4, the limits presented below are much better than those obtained in a recent paper [196].

To estimate the constraints on k_{1q} we will need the proton and neutron expectation values for the quark scalar operators. They are [180, 280, 281]:

$$\langle p|\bar{u}u|p\rangle = 6\bar{p}p, \quad \langle n|\bar{d}d|n\rangle = 6\bar{n}n, \\ \langle p|\bar{d}d|p\rangle = 5\bar{p}p, \quad \langle n|\bar{u}u|n\rangle = 5\bar{n}n. \tag{8.57}$$

Then

$$k_1(\mathrm{Tl}) = 0.4 k_{1p} + 0.6 k_{1n} \approx 5.5 \, (k_{1u} + k_{1d}),$$

and we obtain in this way from (8.33)

$$k_{1u} + k_{1d} < 6 \times 10^{-8}. \tag{8.58}$$

As to k_{2q}, the tensor expectation value can be estimated by dimensional arguments as follows:

$$\langle N|\bar{q}\sigma_{\mu\nu}q|N\rangle \approx \bar{N}\sigma_{\mu\nu}N. \tag{8.59}$$

With this estimate we derive from (8.43)

$$k_{2q} < 10^{-7}. \tag{8.60}$$

Going over to k_{3q}, we have to evaluate the pseudoscalar expectation value. Invariance arguments dictate that

$$\langle N|\bar{q}\gamma_5 q|N\rangle = \rho \bar{N}\gamma_5 N.$$

To obtain the constant ρ in this relation, let us note that

$$\bar{q}\gamma_5 q = \frac{\mathrm{i}}{2m_q}\,\partial_\mu(\bar{q}\gamma_5\gamma_\mu q)$$

and

$$\bar{N}\gamma_5 N = \frac{\mathrm{i}}{2m_\mathrm{p}}\,\partial_\mu(\bar{N}\gamma_5\gamma_\mu N).$$

Now, a simple estimate for the axial current expectation value is

$$\langle N|\bar{q}\gamma_5\gamma_\mu q|N\rangle \approx \bar{N}\gamma_5\gamma_\mu N.$$

The corresponding value of the ratio of interest

$$\rho \approx \frac{m_\mathrm{p}}{m_q}$$

can be substantiated by more refined arguments. Taken together with (8.48) it gives

$$k_{3q} < 10^{-7}. \tag{8.61}$$

Thus, for all three CP-odd electron–quark constants, the upper limits derived from atomic experiments are at the level of 10^{-7}.

In conclusion of this section let us mention that the upper limit (6.5) on $d(\mathrm{Tl})$, being used directly, also seriously constrains the CP-odd electron–electron interaction [282]. It can be easily demonstrated that the effect will be dominated by the interaction of the outer electron with the K-shell electrons (see the corresponding discussion of the P-odd, but T-even electron–electron interaction in Ref. [23], Sect. 4.4). The atomic EDM induced by the CP-odd ee interaction differs from that due to the eN Hamiltonian H_1 for two main reasons. First, the probability density for the outer electron at $r \sim a/Z$ differs from its value at $r \sim R_0$ basically through the absence of the relativistic enhancement factor R. Second, here we deal with an interaction with two K-shell electrons, and not with A nucleons. In this way we obtain the upper limit on the CP-odd electron–electron interaction constant, which is roughly $AR \approx 10^3$ times worse than the bound (8.33):

$$k_s^e < 3 \times 10^{-4}. \tag{8.62}$$

8.4 Electron EDM in Diamagnetic Atoms

Finally, the experimental searches for the EDM of the ground state of the ^{129}Xe atom [242] and of the ^{199}Hg atom [80] also give information on the electron dipole moment. It has been pointed out in [242, 283] that the electron EDM induces a dipole moment of a diamagnetic atom or molecule with closed electron shells due to the hyperfine interaction polarizing these shells along the nuclear spin. The situation is very similar to that with the constants $k_{1p,n}$. Calculations of the electron EDM effect in the diamagnetic atoms were carried out in [265, 275] with the results presented below.

First note that the atomic EDM arising in such a way without including the interaction with the Coulomb field of the nucleus, in second order of perturbation theory,

$$2\sum_n \frac{\langle 0|H_{\rm hf}|n\rangle\langle n|d_{\rm e}\gamma_0\boldsymbol{\Sigma}|0\rangle}{E_0 - E_n},$$

has, as can be easily seen, the order of magnitude $Z\alpha^2(m/m_{\rm p}) \times Z^2\alpha^2 d_{\rm e}$ and is negligibly small.

Much more essential is another mechanism for producing the atomic EDM, the direct interaction of the electron dipole moment with the magnetic moment $\boldsymbol{M}$ of the nucleus. The corresponding noncovariant perturbation can be readily obtained from (2.5):

$$W_{\rm d} = -{\rm i}d_{\rm e}\boldsymbol{\gamma}\cdot\boldsymbol{B}; \tag{8.63}$$

$$\boldsymbol{B} = \boldsymbol{\nabla}\times\left(\boldsymbol{\nabla}\frac{1}{r}\times\boldsymbol{M}\right) = \frac{3(\boldsymbol{M}\cdot\boldsymbol{r})\boldsymbol{r} - \boldsymbol{M}r^2}{r^5} + \frac{8\pi}{3}\boldsymbol{M}\delta(\boldsymbol{r});$$

$$\boldsymbol{M} = \frac{e\mu}{2m_{\rm p}}\frac{\boldsymbol{I}}{I}.$$

The only matrix element of operator (8.63) which is singular in the nuclear radius R_0, is that of mixing of the $s_{1/2}$ and $p_{1/2}$ states:

$$\langle s_{1/2}|W_{\rm d}|p_{1/2}\rangle = \frac{1}{3}\frac{e\mu d_{\rm e}}{m_{\rm p}}\frac{Z^2\alpha R}{a^3(\nu_s\nu_p)^{3/2}}2\boldsymbol{j}\cdot\boldsymbol{I}/I. \tag{8.64}$$

In the nonrelativistic approximation interaction (8.63) reduces to the form

$$W_{\rm d} = -\frac{d_{\rm e}}{2m}\boldsymbol{\sigma}\cdot[(\boldsymbol{p}\times\boldsymbol{B}) - (\boldsymbol{B}\times\boldsymbol{p})].$$

Since in this limit the spin and coordinate variables are decoupled, the sum of these operators for all the electrons is proportional to their total spin, and therefore cannot induce an EDM of closed shells. In other words, in the limit $Z^2\alpha^2 \to 0$ such contributions to the atomic EDM from excitations of the $p_{1/2}$ and $p_{3/2}$ electrons must mutually cancel. Since for the mixing matrix elements $\langle s_{1/2}|W_{\rm d}|p_{3/2}\rangle$ the relativistic enhancement factor is nonsingular in R_0 and therefore close to unity, such a cancellation can be allowed for by the substitution $R \to R - 1$ in the final answer.

As to the atomic EDM induced by the combined action of operator (8.8) and the hyperfine interaction (8.49), it also arises in third order of perturbation theory and has the form

$$\sum_{n,k} \frac{\langle 0| - e\boldsymbol{r}|n\rangle\langle n|V_{\text{hf}}|k\rangle\langle k|V_{\text{d}}|0\rangle}{(E_0 - E_n)(E_0 - E_k)} \tag{8.65}$$

$+$ permutations of operators.

As in the expression (8.50), of the permutations in equation (8.65) one can neglect those where the matrix elements of $-e\boldsymbol{r}$ are placed in the middle, between those of operators V_{hf} and V_{d}. When treating other terms in this expression, it is again convenient to introduce the effective operator

$$\widetilde{V_{\text{d}}} = \sum_k \frac{V_{\text{hf}}|k\rangle\langle k|V_{\text{d}} + V_{\text{d}}|k\rangle\langle k|V_{\text{hf}}}{E_0 - E_k}\,. \tag{8.66}$$

The calculations here are as tedious as in the case of the effective operator (8.51). They lead to the following expression for the only matrix element of $\widetilde{V_{\text{d}}}$ which is singular in the nuclear radius R_0:

$$\langle s_{1/2}|\widetilde{V_{\text{d}}}|p_{1/2}\rangle = -\frac{8}{3}\frac{e\mu d_{\text{e}}}{m_{\text{p}}}\frac{Z^2\alpha R}{a^3(\nu_s\nu_p)^{3/2}}\frac{2\boldsymbol{j}\cdot\boldsymbol{I}}{I}. \tag{8.67}$$

Now, in the nonrelativistic limit the sum of operators V_{d} of all the electrons is proportional to their total spin which is equal to zero both in the ground state $|0\rangle$ and in the state $\boldsymbol{r}\,|0\rangle$. As a result, in this limit the initial expression (8.65) for the atomic EDM vanishes. In other words, it means, as in the previous case, that in the nonrelativistic approximation there is a cancelation of the contributions of the excitations of the $p_{1/2}$ and $p_{3/2}$ electrons. And again this cancellation is taken into account through the substitution $R \to R-1$ in the answer.

Let us compare now the resulting matrix element

$$\langle s_{1/2}|W_{\text{d}} + \widetilde{V_{\text{d}}}|p_{1/2}\rangle \to -\frac{7}{3}\frac{e\mu d_{\text{e}}}{m_{\text{p}}}\frac{Z^2\alpha\,(R-1)}{a^3(\nu_s\nu_p)^{3/2}}\frac{2\boldsymbol{j}\cdot\boldsymbol{I}}{I} \tag{8.68}$$

with the tensor contribution to (8.25). In this way we arrive at the correspondence

$$d_{\text{e}}\frac{\boldsymbol{I}}{I} \to \frac{3}{7}\frac{Gm_{\text{p}}e}{\sqrt{2}\pi\alpha\mu}\frac{R}{R-1}\left\langle k_{2p}\sum_p \boldsymbol{\sigma}_p + k_{2n}\sum_n \boldsymbol{\sigma}_n\right\rangle. \tag{8.69}$$

The accuracy of this relation is $\sim Z^2\alpha^2$. The upper limits on the electron EDM following in this way from the xenon and mercury experiments are, respectively,

$$d_{\text{e}} < 2 \times 10^{-23}\ e\,\text{cm} \tag{8.70}$$

and

$$d_{\text{e}} < 8 \times 10^{-26}\ e\,\text{cm}. \tag{8.71}$$

For xenon the results of [275] and [265] are in good agreement. Unfortunately, in the case of mercury the Hartree–Fock result of [265] differs greatly (by a factor close to -0.8) from the phenomenological one (8.71) obtained in [275]. The reason for such a serious discrepancy is not clear. The authors of [265] point out, however, that higher-order corrections in mercury are much more important than in xenon, and correlation effects of second and higher order may change their result.

8.5 CP-Odd Nuclear Moments

8.5.1 The Schiff Moment

The most important information on the CP violation in the hadronic sector, up to now obtained from atomic experiments, is due to the effect of the finite size of the nucleus. Of course, in principle, not only the electrostatic forces are essential here, but the nuclear forces as well. However, the correction to the commutator in (6.2) resulting from the strong interactions depends only on the nucleon coordinates and hence leads by itself to no mixing of the electron states in the atom. Nonetheless, as has been shown by Schiff [200], the nuclear EDM induces an atomic dipole moment if the charge and EDM distributions in the nucleus are different. Indeed, for noncoincident densities of nuclear charge and dipole moment, ρ_q and ρ_d, the EDM interaction

$$\int \mathrm{d}\boldsymbol{r}'\, \rho_d(\boldsymbol{r}')\, \boldsymbol{d}_N \boldsymbol{\nabla}' A_0(\boldsymbol{r} - \boldsymbol{r}')$$

does not reduce to the commutator with the Hamiltonian in which the potential energy is

$$Ze \int \mathrm{d}\boldsymbol{r}'\, \rho_q(\boldsymbol{r}')\, A_0(\boldsymbol{r} - \boldsymbol{r}').$$

It can now be easily demonstrated that the interaction of the electron with the dipole moment of a finite-size nucleus, resulting in an atomic EDM, can be written as

$$V_s = \int \mathrm{d}\boldsymbol{r}'\, [\rho_d(\boldsymbol{r}') - \rho_q(\boldsymbol{r}')]\, \boldsymbol{d}_\mathrm{N} \boldsymbol{\nabla}' \frac{-e}{|\boldsymbol{r} - \boldsymbol{r}'|}. \tag{8.72}$$

Here $\boldsymbol{r}$ is the electron coordinate measured from the center of the nucleus, and the densities ρ_q and ρ_d are normalized by the condition $\int \mathrm{d}\boldsymbol{r}\rho_{q,d} = 1$. Limiting ourselves to the first nonvanishing term of the expansion in r'/r, we transform the above expression to

$$V_s = \frac{1}{2} e \int \mathrm{d}\boldsymbol{r}'\, [\rho_d(\boldsymbol{r}') - \rho_q(\boldsymbol{r}')]\, d_{\mathrm{N},l}\, r'_m r'_n\, \nabla_l \nabla_m \nabla_n \frac{1}{r}. \tag{8.73}$$

The charge density is obviously spherically symmetric up to corrections of order Z^{-1}. Even for deformed nuclei, where these corrections may be an order of magnitude greater, they can still be neglected usually (see, however, Sect. 10.3). Thus the dipole moment of the nucleus will be assumed here to coincide with the EDM of the valence nucleon, so that $\boldsymbol{d}_{\rm N} = d_{\rm p,n}\boldsymbol{\sigma}$, and ρ_d coincides with the density of the valence nucleon distribution. In this case expression (8.73) can be rewritten as

$$V_s = \frac{1}{2}\, e\, d_{\rm p,n} \int {\rm d}r'\, 4\pi (r')^2 \Big[\rho_d(\boldsymbol{r}\,')\langle \sigma_l n_m n_n\rangle - \rho_q(r')\, \frac{1}{3}\, \delta_{mn}\langle\sigma_l\rangle\Big]\, \nabla_l \nabla_m \nabla_n\, \frac{1}{r}. \tag{8.74}$$

Here $\boldsymbol{n} = \boldsymbol{r}\,'/r'$ and the brackets $\langle ... \rangle$ denote averaging over the state with a given angular momentum $\boldsymbol{I}$ of the nucleus.

Let us split the "electron" factor $\nabla_l \nabla_m \nabla_n\, 1/r$ in the interaction (8.74) into two terms

$$\Big[\nabla_l \nabla_m \nabla_n - \frac{1}{5}\,(\delta_{lm}\nabla_n + \delta_{mn}\nabla_n + \delta_{nl}\nabla_m)\Delta\Big]\, \frac{1}{r} + \frac{1}{5}\,(\delta_{lm}\nabla_n + \delta_{mn}\nabla_n + \delta_{nl}\nabla_m)\Delta\, \frac{1}{r}.$$

The first term, an irreducible third-rank tensor, corresponds to the interaction of the electron with the electric octupole moment of the nucleus. It results in mixing of only the atomic states with large angular momenta, so that its contribution to the atomic EDM is strongly suppressed. In any case, an EDM of the ground state of the atoms that are of experimental interest, cesium, thallium, xenon and mercury, is not induced by the octupole.

As for the second term, its contribution to the Hamiltonian (8.74) can be expressed as

$$\begin{aligned} V_s &= -\frac{1}{2}\, e\, d_{\rm p,n} \Big[r_q^2\, \frac{1}{3}\Big(K - \frac{1}{2}\Big) - r_d^2\, \frac{1}{5}\Big(K - \frac{3}{2}\Big)\Big]\, \frac{\boldsymbol{I}\boldsymbol{\nabla}}{I(I+1)}\, 4\pi\delta(\boldsymbol{r}) \\ &= -e\boldsymbol{S}\boldsymbol{\nabla}\, \delta(\boldsymbol{r}). \end{aligned} \tag{8.75}$$

Here $K = (l - I)(2I + 1)$ and l is the orbital angular momentum of the valence nucleon; the vector coefficient $\boldsymbol{S}$ in this expression will be called the Schiff moment (SM) of the nucleus.

What can be said about the mean squared radii $r^2_{q,d} = \int {\rm d}\boldsymbol{r}\,'(r')^2\rho_{q,d}(\boldsymbol{r}\,')$? It is known, for instance from the scattering of electrons on nuclei, that the nuclear charge can be considered to good accuracy to be uniformly distributed over a sphere of radius $R_0 = 1.2\, A^{1/3}$ fm. Then $r_q^2 = (3/5)R_0^2$. For the magnitude of r_d^2, it would be natural to assume that it coincides with the mean squared magnetic radius of the nucleus r_m^2. The latter quantity has been measured for tritium, ^{3}He, ^{45}Sc, ^{51}V, and ^{59}Co. Since for all these nuclei, except for ^{3}He, there is no significant difference between r_q^2 and r_d^2 (see Table 8.2), we can assume $r_d^2 = r_q^2$. As a result, the expression for the Schiff moment simplifies to

Table 8.2. Charge and magnetic radii of nuclei

	r_q^2 [fm^2]	r_m^2 [fm^2]
$^3H_1$	2.82 ± 0.34 [284]	2.66 ± 0.33 [284]
$^3He_2$	3.88 ± 0.40 [284]	2.86 ± 0.34 [284]
$^{45}Sc_{21}$	12.39 ± 0.63 [285]	12.67 ± 1.57 [286]
$^{51}V_{23}$	13.10 ± 0.65 [285]	12.89 ± 0.72 [286]
$^{59}Co_{27}$	14.21 ± 0.38 [285]	13.99 ± 1.05 [286]

$$\boldsymbol{S} = d_{\mathrm{p,n}} R_0^2 \frac{4\pi}{25} \frac{(K+1)\boldsymbol{I}}{I(I+1)}. \tag{8.76}$$

It follows from formula (8.74) that at $r_d^2 = r_q^2$ the effect under discussion is proportional to $\langle n_m n_n - (1/3)\delta_{mn}\rangle$ and is therefore related to the orbital quadrupole moment of the nucleus.

In spite of being local, the SM interaction (8.75) is singular enough to induce not only $s_{1/2} - p_{1/2}$ mixing in atoms, but $s_{1/2} - p_{3/2}$ as well. The mixing matrix element is

$$\langle s_{1/2}|V_s|p_j\rangle = \frac{eZ^2R_j}{a^4(\nu_s\nu_p)^{3/2}} \boldsymbol{S}\,\langle \Omega_{1/20}|\boldsymbol{n}|\Omega_{j1}\rangle. \tag{8.77}$$

The relativistic enhancement factors R_j are different for $p_{1/2}$ and $p_{3/2}$ electrons:

$$R_{1/2} = \frac{3\gamma_{1/2}}{2\gamma_{1/2}+1} R \approx R, \tag{8.78}$$

$$R_{3/2} = \frac{6[(\gamma_{1/2}+1)(\gamma_{3/2}+2) + Z^2\alpha^2]}{\Gamma(2\gamma_{1/2}+1)\,\Gamma(2\gamma_{3/2}+1)} \times(\gamma_{1/2}+\gamma_{3/2}-2)\left(\frac{a}{2ZR_0}\right)^{3-\gamma_{1/2}-\gamma_{3/2}} \tag{8.79}$$

Here $\gamma_j = \sqrt{(j+1/2)^2 - Z^2\alpha^2}$. The calculation itself differs from that leading to expression (8.25) in only one respect: one has to include one more term of the Bessel function expansion.

For ^{133}Cs ($I = 7/2,\ K = 4$) the interaction (8.75) reduces to

$$V_s = -\frac{16\pi\, ed_{\mathrm{p}}}{315} R_0^2\,(\boldsymbol{I}\boldsymbol{\nabla})\delta(\boldsymbol{r}). \tag{8.80}$$

Considering only the admixture of the $6p$ states to the cesium ground state $6s_{1/2}$, we arrive at the following expression for the ratio of the atomic EDM to that of the proton [287]:

$$K_s(\mathrm{Cs}) = \frac{Z^2 R_0^2}{a^2} \frac{(R_{1/2} + 2R_{3/2})\,\rho(6s, 6p)}{(\nu_{6s}\nu_{6p})^{3/2}} \frac{\mathrm{Ry}}{E(6s) - E(6p)} \times \begin{cases} 32/405, & F = 4 \\ 8/105, & F = 3\,. \end{cases} \tag{8.81}$$

To simplify this expression, we have neglected the small difference between the $6p_{1/2}$ and $6p_{3/2}$ states in the energies and in the E1 radial integrals. As to the analysis of the corresponding prediction for the proton EDM, it is more convenient to delay it to the next section.

We move on now to thallium. According to the shell model, the valence proton in both stable isotopes, ^{203}Tl and ^{205}Tl, is in the $s_{1/2}$ state. Thus, the interaction (8.75) reduces to the form

$$V_s = \frac{4\pi}{3}\, ed_{\mathrm{p}}\, \delta r^2\, (\boldsymbol{I}\boldsymbol{\nabla})\delta(\boldsymbol{r}), \quad \delta r^2 = r_q^2 - r_d^2. \tag{8.82}$$

In this case the ratio of the atomic and proton EDMs amounts to [287]

$$K_s(\mathrm{Tl}) \approx 1.3 \times 10^{-3} \frac{\delta r^2}{R_0^2}. \tag{8.83}$$

Unfortunately, a reliable calculation of δr^2 is quite difficult due simply to the above-mentioned closeness of the values of r_q^2 and r_d^2. Relatively realistic calculations with the Woods–Saxon potential result for thallium in $\delta r^2/R_0^2 \approx 0.08$ [278]; for ^{45}Sc, ^{51}V, and ^{59}Co, where the proton is in the $f_{7/2}$ state, the same calculations give $\delta r^2/R_0^2 \approx -0.08$ which at least does not contradict the experimental numbers (see Table 8.1). However, according to [288], δr^2 is very sensitive to nuclear many-body effects. When the latter are allowed for in thallium, the result becomes $\delta r^2/R_0^2 = -0.04 \pm 0.08$.

The uncertainty in δr^2 hampers the extraction of reliable information on the proton EDM from the thallium experiment. If we tentatively put $\delta r^2/R_0^2 = 0.08$, the constraint on the proton dipole moment following from the thallium result (6.5) would be

$$d_{\mathrm{p}} < 2.4 \times 10^{-20}\, e\,\mathrm{cm}. \tag{8.84}$$

A better limit on the proton EDM can be derived [274] from the experiment with ^{129}Xe in the ground state. The atomic EDM, as induced by the nuclear Schiff moment, in this case equals

$$d(^{129}\mathrm{Xe}) = \frac{8}{9\pi} S \frac{Z^2}{a^2} (R_{1/2} + 2R_{3/2}) \times \sum_n \frac{\rho(5p, ns)}{(\nu_{5p}\nu_{ns})^{3/2}} \frac{\mathrm{Ry}}{E(5p) - E(ns)}. \tag{8.85}$$

This formula can be derived in practically the same way as (8.81) for cesium, the only real difference being that now we cannot restrict ourselves to intermediate states of a single n. In particular, we neglect again the difference between the $5p_{1/2}$ and $5p_{3/2}$ orbitals in the energies and in the E1 radial integrals. Here the origin of different coefficients of $R_{1/2}$ and $R_{3/2}$ becomes especially obvious: the contribution of four $5p_{3/2}$ electrons to the effect is twice as large as that of two $5p_{1/2}$ electrons.

Still, we would like to take this opportunity to demonstrate a technical trick which allows one to single out explicitly the contribution of the excitations $5p_{3/2} - ns_{1/2}$ without any Clebsch–Gordan coefficients. Such a contribution for a given n is proportional to

$$\sum_{j_z, j_{z'}} \langle ns_{1/2}, j_{z'} | \langle I, I_z | \\ \times \int d\boldsymbol{r}\, (\boldsymbol{S}\boldsymbol{\nabla})\delta(\boldsymbol{r}) | I, I_z \rangle | \, 5p_{3/2}, j_z \rangle \langle 5p_{3/2}, j_z | \, z \, | \, ns_{1/2}, j_{z'} \rangle .$$

We have taken into account here that the dipole moment operator $-e\boldsymbol{r}$ does not change the nuclear spin projection I_z. The sum over j_z at fixed $j = 3/2$ can be conveniently transformed, using the projection operator

$$P_{3/2} = \frac{1}{3}\left[\left(\boldsymbol{l} + \frac{1}{2}\boldsymbol{\sigma}\right)^2 - \frac{3}{4}\right] = \frac{1}{3}(\boldsymbol{l}\cdot\boldsymbol{\sigma} + 2),$$

into a sum over all possible projections of the orbital angular momentum $l = 1$ and spin $s = 1/2$. Then, taking into account that the state $z \, | \, ns \rangle$ is expanded only in functions with $l = 1$, we can formally go over to summation over l as well. The resulting extended sum is nothing but the completeness relation and is thus trivially calculated. The summation over $j_{z'}$, as can be easily seen, then reduces to a trace of the spin operator thus obtained.

Though for the contribution of the excitations $5p_{1/2} - ns_{1/2}$ an analogous simplification is less crucial, in this case it would also be convenient, without using the representation (8.2) for the angular part of the wave function $p_{1/2}$, to transform an analogous sum over j_z to the completeness relation using the projection operator

$$P_{1/2} = \frac{1}{3}(-\boldsymbol{l}\cdot\boldsymbol{\sigma} + 1).$$

A simple estimate by formula (8.85) under the assumptions used in Sect. 8.3, $\rho \sim 1$, $E(ns) \sim -E(5p)$, $\nu_s \sim \nu_p \sim 1$, gives

$$|d(^{129}\text{Xe})| \sim 10^{20} S \cdot \text{cm}^{-2}.$$

The above-mentioned Hartree–Fock calculations [273, 274] result, respectively, in

$$d(^{129}\text{Xe}) = 2.7 \times 10^{20} S \cdot \text{cm}^{-2} \tag{8.86}$$

and

$$d(^{129}\text{Xe}) = 2.1 \times 10^{20} S \cdot \text{cm}^{-2}. \tag{8.87}$$

Comparing these relations with the experimental limit (7.3), we find [1]

[1] The quantity Q used in [274, 275, 278] is related to S through $Q = S/4\pi$.

$$S(^{129}\mathrm{Xe}) < 10^{-7}\, e\,\mathrm{fm}^3. \tag{8.88}$$

Then, the Schiff moment $S(^{129}\mathrm{Xe})$ caused by the EDM of the unpaired nucleons can, according to the considerations given in Section 8.3, be represented as

$$S(^{129}\mathrm{Xe}) = -\frac{2\pi}{3}\,\delta r^2\,(0.24 d_\mathrm{p} + 0.76 d_\mathrm{n}). \tag{8.89}$$

The calculation [274] with the Woods–Saxon potential gives for the $^{129}\mathrm{Xe}$ nucleus $\delta r^2 = -3.8\ \mathrm{fm}^3$. As for the neutron EDM, d_n, the direct upper limit (4.43) is well below the level discussed here. Thus from the comparison of (8.88) and (8.89) it follows that [274]

$$d_\mathrm{p} < 5 \times 10^{-21}\, e\,\mathrm{cm}. \tag{8.90}$$

However modest the results (8.84) and (8.90) might appear compared with the upper limits on the neutron and electron EDMs, they are five orders of magnitude better than the upper limit following from the measurements of the spin precession of a free proton [232].

But more interesting results are obtained from the mercury experiment [80]. Unfortunately, there is no direct calculation of the SM contribution to the $^{199}\mathrm{Hg}$ dipole moment. However, it can be found [289] from the relations (8.38), (8.42) and (8.86), originally obtained in [273]. To this end one should multiply relation (8.86) by the ratio of the numerical factors in (8.42) (which should first be rewritten as $\boldsymbol{d}(^{199}\mathrm{Hg}) = -6.0 \times 10^{-20}\, k_{2n}\langle\boldsymbol{\sigma}\rangle\ e\,\mathrm{cm}$) and in (8.38). The result is

$$d(^{199}\mathrm{Hg}) = -3.1 \times 10^{21}\, S \cdot \mathrm{cm}^{-2}. \tag{8.91}$$

Now from the experimental upper limit (7.4) we obtain

$$S(^{199}\mathrm{Hg}) < 2.8 \times 10^{-10}\, e\,\mathrm{fm}^3. \tag{8.92}$$

Finally, relating the Schiff moment $S(^{199}\mathrm{Hg})$ to the neutron EDM, d_n, through formula (8.76), we arrive at the following upper limit [290]:

$$d_\mathrm{n} < 8 \times 10^{-25}\, e\,\mathrm{cm}. \tag{8.93}$$

This is not so different from the best results (4.43) of the direct measurements.

Unfortunately, the theoretical uncertainty in this result is considerable. The accuracy of the atomic many-body calculation for mercury can be estimated perhaps as 30%. The error introduced by using the simple nuclear shell model may be as high: for instance, the model prediction for the magnetic moment $\mu(^{199}\mathrm{Hg})$ is 0.64 while the experimental value is 0.5.

8.5.2 Magnetic Quadrupole Moment

The Schiff moment is not the only way to circumvent the screening of nuclear dipole moments in atoms. One more effect is the interaction of the electron with the magnetic quadrupole moment (MQM) of the nucleus [287]. The physical origin of the MQM is clear. While the orbital motion of a charged particle results in the appearance of a magnetic dipole moment for the system, the orbital motion of a particle with an EDM gives rise to a magnetic quadrupole moment.

Let us calculate this effect. Variation with respect to the vector potential $\boldsymbol{A}$ (see [15], §115) of expression (2.5) for the interaction of the nucleon dipole moment d_{N} with an external field gives the following contribution to the current density

$$\boldsymbol{j}_{\mathrm{d}} = \mathrm{i}\, d_{\mathrm{N}}\, \boldsymbol{\nabla} \times (u'^{\dagger} \boldsymbol{\gamma}\, u).$$

Going over now from the four-component relativistic wave functions u to the usual two-component spinors ψ and limiting ourselves to first-order terms in m_{p}^{-1}, we obtain

$$\boldsymbol{j}_{\mathrm{d}} = \frac{d_{\mathrm{N}}}{2m_{\mathrm{p}}}\, \boldsymbol{\nabla} \times [\,\psi'^{\dagger} \boldsymbol{\sigma} \times (\boldsymbol{p}' + \boldsymbol{p})\, \psi\,]. \tag{8.94}$$

Here $\boldsymbol{p}$ and $\boldsymbol{p}'$ are the momentum operators acting on ψ and $\psi'^{\dagger}$, respectively. Let us simplify the current operator using the shell model of the nucleus. (As mentioned above, the theoretical ambiguity introduced in this way is rather high; in cesium, discussed in detail below, it can reach 30%.) We begin by averaging the operator

$$\frac{1}{2m_{\mathrm{p}}}\, (\boldsymbol{p}' + \boldsymbol{p})$$

over the state of the valence nucleon with a given orbital angular momentum $\boldsymbol{l}$. In fact, we consider the current inducing the orbital magnetic moment of the nucleus. Assuming the nucleus to be pointlike, we can represent the corresponding magnetic field as

$$\boldsymbol{B} = \frac{e}{2m_{\mathrm{p}}}\, \boldsymbol{\nabla} \times \left(\boldsymbol{\nabla} \frac{1}{r} \times \boldsymbol{l} \right).$$

Using the Maxwell equation $\boldsymbol{\nabla} \times \boldsymbol{B} = 4\pi \boldsymbol{j}$, we find, after simple transformations,

$$\frac{1}{2m_{\mathrm{p}}} \overline{(\boldsymbol{p}' + \boldsymbol{p})} = \frac{1}{2m_{\mathrm{p}}}\, \boldsymbol{\nabla}\, \delta(\boldsymbol{r}) \times \boldsymbol{l}.$$

From this we get

$$\boldsymbol{j}_{\mathrm{d}} = -\frac{d_{\mathrm{N}}}{2m_{\mathrm{p}}}\, \boldsymbol{\nabla} \times \overline{[\,\boldsymbol{l}\,(\boldsymbol{\sigma} \cdot \boldsymbol{\nabla})\, \delta(\boldsymbol{r})\,]}.$$

Here the bar denotes averaging over the state of the nucleus with a given total angular momentum $\boldsymbol{I}$. The tensor $\overline{l_m \sigma_n}$ is easily shown to be symmetric, and the term containing δ_{mn} makes no contribution to $\boldsymbol{j}_{\mathrm{d}}$. Using all the above results, we get

$$j_{\mathrm{d},m} = -\frac{d_{\mathrm{N}}}{2m_{\mathrm{p}}}\frac{1/2-K}{2I(I+1)}\epsilon_{mrn}\left[I_rI_s+I_sI_r-\frac{2}{3}\delta_{rs}I(I+1)\right]\nabla_s\nabla_n\,\delta(\boldsymbol{r}).$$

The vector potential induced by this current is

$$A_m = \frac{d_{\mathrm{N}}}{2m_{\mathrm{p}}}\frac{1/2-K}{2I(I+1)}\epsilon_{mrn}\left[I_rI_s+I_sI_r-\frac{2}{3}\delta_{rs}I(I+1)\right]\nabla_s\nabla_n\,\frac{1}{r}. \quad (8.95)$$

We will call the tensor

$$M_{mn} = \frac{3}{2}\frac{d_{\mathrm{N}}}{m_{\mathrm{p}}}\frac{1/2-K}{I(I+1)}\left[I_mI_n+I_nI_m-\frac{2}{3}\delta_{mn}I(I+1)\right] \quad (8.96)$$

the operator of the nuclear magnetic quadrupole moment. As usual, we will characterize the nuclear MQM by the expectation value of M_{zz} in the state with the maximum I_z:

$$\mathcal{M} = \frac{d_{\mathrm{N}}}{m_{\mathrm{p}}}\frac{(2I-1)(1/2-K)}{I+1}. \quad (8.97)$$

It can be easily demonstrated that definitions (8.96), (8.97) are in accordance with formulas (2.9) and (2.10).

For those who find the above derivation of the expressions for the nuclear MQM and the vector potential induced by it too intricate, the same relations can be derived by using the standard angular momentum technique.

The interaction of a relativistic electron with the vector potential (8.95) expressed directly in terms of $\mathcal{M}$ is

$$V_{\mathrm{M}} = -\frac{e\mathcal{M}}{4I(2I-1)}\left[I_mI_n+I_nI_m-\frac{2}{3}\delta_{mn}I(I+1)\right] \times\frac{1}{2}\alpha_k\nabla_l(\epsilon_{klm}\nabla_n+\epsilon_{kln}\nabla_m)\frac{1}{r}. \quad (8.98)$$

This interaction results in mixing of opposite-parity levels with the same total angular momentum of the atom F. Let us note that, being an irreducible second-rank tensor in the electron variables, the interaction (8.98) does not mix $s_{1/2}$ and $p_{1/2}$ states. The matrix element of the operator

$$\frac{1}{2}\alpha_k\nabla_l(\epsilon_{klm}\nabla_n+\epsilon_{kln}\nabla_m)\frac{1}{r}$$

will be written as

$$\frac{3}{2}\langle j_2\,|\,\sigma_m n_n+\sigma_n n_m-2\,(\boldsymbol{\sigma}\cdot\boldsymbol{n})\,n_m n_n\,|\,j_1\rangle\,P,$$

where the radial integral P can be calculated using the semiclassical wave functions (8.3):

$$P = \int_0^\infty \frac{\mathrm{d}r}{r}\,(f_2g_1+f_1g_2) = \frac{Z^2\alpha}{a^3(\nu_1\nu_2)^{3/2}}\,\lambda\,R_{\mathrm{M}}. \quad (8.99)$$

Here

$$\lambda = \frac{\kappa_1\kappa_2}{|\kappa_1\kappa_2|}\,\frac{96\,(\kappa_1+\kappa_2-2)\,(j_1+j_2-2)!}{(j_1-j_2+2)!\,(-j_1+j_2+2)!\,(j_1+j_2+3)!},$$

and the relativistic enhancement factor $R_{\rm M}$ is introduced in the following way:

$$R_{\rm M} = \frac{(j_1-j_2+2)!\,(-j_1+j_2+2)!\,(j_1+j_2+3)!\,\Gamma(\gamma_1+\gamma_2-2)}{(j_1+j_2-2)!\,\Gamma(\gamma_1-\gamma_2+3)\,\Gamma(-\gamma_1+\gamma_2+3)\,\Gamma(\gamma_1+\gamma_2+3)}.$$

It should be noted that in the case of mixing of states of higher orbital angular momenta than s and p, this expression for the radial integral is only an estimate. First, the contribution of short distances loses its dominance with growing l. Second, the role of many-body effects increases.

Obviously, the MQM arises only in nuclei with $I > 1/2$. Besides, the nuclear MQM induces the dipole moment only in atoms with nonvanishing electron angular momentum since the magnetic quadrupole moment interacts directly with the magnetic field of the electrons. Therefore, among the atoms which up to now have been objects of experiments, it is only cesium where the nuclear MQM is operative.

When calculating the EDM of the ground state of the cesium atom, it is sufficient to restrict oneself to admixtures of the $6p$ states, in this case of the $6p_{3/2}$ state. Here $\lambda = -2/3$, $R_{\rm M} = 1.3$ and the mixing matrix element is

$$\langle p_{3/2}\,|\,V_{\rm M}\,|\,s_{1/2}\rangle = -\frac{d_{\rm p}}{ea}\,\frac{m}{m_{\rm p}}\,\frac{Z^2\alpha^2 R_{\rm M}{\rm Ry}}{(\nu_{6s}\nu_{6p})^{3/2}}\,\frac{4}{3} \times \begin{cases} \sqrt{7/5}, & F=4 \\ -\sqrt{3}, & F=3\,. \end{cases} \tag{8.100}$$

The experimental values of the required parameters are [259, 260]: the energy interval $E(6p_{3/2}) - E(6s) = 11732\,{\rm cm}^{-1}$; the effective principal quantum numbers $\nu(6s) = 1.87$, $\nu(6p_{3/2}) = 2.37$; the radial integral $\rho(6s, 6p_{3/2}) = -5.509(7)$. So the ratio of the atomic dipole moment, as induced through the nuclear MQM, to the proton EDM amounts to

$$\begin{aligned} K_{\rm M}({\rm Cs}) &= \frac{m}{m_{\rm p}}\,\frac{Z^2\alpha^2 R_{\rm M}\rho(6s,\,6p_{3/2})}{(\nu_{6s}\nu_{6p})^{3/2}} \\ &\quad\times \frac{\rm Ry}{E(6p_{3/2})-E(6s)} \times \begin{cases} -56/15, F-4 \\ \quad 2, \quad F=3 \end{cases} \\ &= 10^{-3} \times \begin{cases} 0.78, \; F=4 \\ -1.26, F=3\,. \end{cases} \end{aligned} \tag{8.101}$$

Let us now compare the contributions to the atomic dipole moment from the nuclear SM and MQM:

$$\begin{aligned} \frac{K_{\rm S}({\rm Cs})}{K_{\rm M}({\rm Cs})} &= m\,m_{\rm p}\,r_0^2\,A^{2/3}\,\frac{R_{1/2}+2\,R_{3/2}}{R_{\rm M}} \times \begin{cases} -4/63, F=4 \\ 4/105, \; F=3 \end{cases} \\ &= \begin{cases} -0.14, \; F=4 \\ \;\; 0.09, F=3\,. \end{cases} \end{aligned} \tag{8.102}$$

So, the quadrupole contribution is the dominant one. This situation is quite common (if MQM is operative at all). Even in the heaviest atoms, the Schiff interaction, despite an obvious growth $\sim A^{2/3}$ and a steeper dependence of $R_{1/2,\,3/2}$ on Z compared with R_M, does not exceed the magnetic quadrupole effect.

The final result for the coefficient K_p in the cesium atom is

$$K_p(\mathrm{Cs}) \;=\; K_\mathrm{M}(\mathrm{Cs}) + K_\mathrm{S}(\mathrm{Cs}) \;=\; 10^{-3} \times \begin{cases} \;\;0.67, & F = 4 \\ -1.37, & F = 3\,. \end{cases} \tag{8.103}$$

Finally, from the experimental result (6.4) obtained for a state with $F = 4$, and from this value of $K_p(\mathrm{Cs})$ we derive the following upper limit on the proton EDM:

$$d_\mathrm{p} < 2 \times 10^{-20}\; e\,\mathrm{cm}. \tag{8.104}$$

On the other hand, comparing (8.102) with the same experimental result (6.4), one can also obtain a limit on the magnetic quadrupole moment of the ^{133}Cs nucleus:

$$\mathcal{M}(\mathrm{Cs}) \;=\; \frac{14}{3}\frac{d_\mathrm{p}}{m_\mathrm{p}} < 1.3 \times 10^{-6}\,\frac{e}{2m_\mathrm{p}}\,r_0; \quad r_0 = 1.2\,\mathrm{fm} \tag{8.105}$$

9. T Violation in Molecules

You boil it in sawdust: you salt it in glue:
You condense it with locusts and tape:
Still keeping one principal object in view –
To preserve its symmetrical shape.
Lewis Carroll

9.1 Enhancement of an Applied Field by a Polar Molecule

As was first addressed by Sandars, the internal energy states of heavy polar molecules can show evidence of EDMs or T violating interactions among the constituents [291]. When a polar molecule is oriented by the application of an external electric field, the mean intermolecular field acting on the nuclei and electrons of the molecule differs from zero. The magnitude of this internal field is enormous, of an atomic scale; for the Tl^+F^- polar molecule, this field amounts to about 10^9 V/cm.

In order for an applied field to significantly orient a polar molecule, the energy given by $\boldsymbol{E} \cdot \boldsymbol{d}$ where $\boldsymbol{d}$ is the molecular dipole moment, must be larger than the rotational level spacings of the molecule, i.e.,

$$E > \frac{B(L+1)(L+2) - BL(L+1)}{d} = \frac{2B(L+1)}{d}, \tag{9.1}$$

where $B = 1/(2MR^2) = 6.7$ GHz is the rotational energy constant, M is the reduced mass of the nuclei, and R is the distance between them. For TlF, with $d = 4.2$ Debye $= 8.7 \times 10^{-9}$ $e\,$cm, this requires for $L = 1$ about 10 kV/cm. Indeed, for a field of about 25 kV/cm, the TlF molecule alignment is close to 76%. We thus see an applied laboratory field enhanced by a factor of order 10^5 within the molecule.

If the molecules are hot, the rotational states will be distributed around a rather large L; only those molecules with small L are effective for an EDM experiment. Various experimental schemes have been used or proposed to rotationally cool molecular beams (supersonic expansion, expansion with an inert carrier gas, laser cooling). But in general the main problem associated with population of the high L rotational states is a loss in statistical accuracy.

The enhancement of P- and T-odd effects in polar molecules can be understood from another point of view [292, 293]: this relates to the very small interval between levels of opposite parity. (By the parity of a level we mean here, as usual, the behavior of its wave function under reflection of the coordinates of all the particles: electrons and nuclei. In molecular spectroscopy the P-even states are called positive, and the P-odd ones negative.) The nature of this new enhancement mechanism, as compared to the case of heavy atoms, is easily understood (its existence was first noted in [294] in connection with P-odd, T-even interactions). In the rest frame of a diatomic molecule, the Hamiltonian possesses axial symmetry and the electron state is therefore characterized by the projection of its angular momentum on the molecular axis $\boldsymbol{\lambda}$ (see [15], §§78, 82). In the situation under discussion, when at least one of the atoms in the molecule is heavy, the spin–orbit interaction is quite large, so that we are dealing with the projection Ω of the total angular momentum of the electron $\boldsymbol{j}$. On reflection in a plane passing through the axis $\boldsymbol{\lambda}$ (let us note that a further rotation through the angle π around the normal to the reflection plane results in inversion of all coordinates) the molecular energy obviously does not change, while the sign of $\Omega = \boldsymbol{j} \cdot \boldsymbol{\lambda}$ reverses. It is therefore clear that the states $|\Omega\rangle$ and $|-\Omega\rangle$ are degenerate. States of definite parity are constructed as linear combinations of $|\Omega\rangle$ and $|-\Omega\rangle$ and, naturally, they are also degenerate. However, the interaction of the electron with the rotation of the molecule results in mixing of the states $|\Omega\rangle$ and $|-\Omega\rangle$ and thus lifts the degeneracy of the opposite parity levels. Such an interaction arises from the rotational energy of the nuclei

$$H_{\rm rot} = B\boldsymbol{L}^2, \quad \boldsymbol{L} = \boldsymbol{J} - \boldsymbol{j}.$$

Here $\boldsymbol{J}$ is the total angular momentum of the molecule, and $\boldsymbol{L}$ is the angular momentum of the molecular rotation. So, the above-mentioned interaction is

$$V = -2B\, \boldsymbol{J} \cdot \boldsymbol{j}. \tag{9.2}$$

Being a vector in the electron variables (for fixed $\boldsymbol{J}$), this perturbation can change the projection of the electron angular momentum only by unity. Therefore, mixing of the levels $|\Omega\rangle$ and $|-\Omega\rangle$ arises only in the $2|\Omega|$th order of the operator (9.2). For $|\Omega| = 1/2$, the mixing matrix element is

$$\langle \Omega = \pm\frac{1}{2}\,|\,\hat{V}\,|\,\Omega = \mp\frac{1}{2}\rangle,$$

and hence the splitting of the levels of opposite parity for $|\Omega| = 1/2$ also has, as is seen from (9.2), the following order of magnitude

$$\Delta E \sim BJ \sim \frac{m}{M}\, J\,\mathrm{Ry}.$$

It is indeed much less than the normal atomic energy interval $E \sim$ Ry, even at large J. (Let us note that otherwise the adiabatic approximation, in which the interaction of the electron with the molecular rotation is considered as a perturbation, would be invalid.) It is exactly because this splitting and the rotational splitting [see (9.1)] are so small, that relatively moderate external fields are sufficient to create quantum-mechanical molecular states which are superpositions of opposite-parity states with comparable weights. In the semiclassical, visual picture this corresponds to an aligned molecule.

It is far from being trivial to incorporate the exact screening of the electric field at the nucleus into intricate molecular calculations. Returning to the example of TlF, there are indications that the effective internal field at the Tl nucleus may be an order of magnitude smaller than not only simple estimates, but even the more serious calculations.

The Tl^+ ion by itself is quite similar to the Hg atom; both have 80 electrons, with the same electronic configuration. Part of the discussion concerning the effects of a nuclear EDM on the Hg atom is directly applicable to the Tl^+. One difference is that the Tl nucleus (both the 203 and 205 isotopes) has a valence proton as compared to a valence neutron in ^{199}Hg. But in the TlF molecule Tl^+ is strongly polarized; that is, atomic p orbitals are strongly admixed to the s orbitals (see below).

TlF is a diamagnetic molecule, and as in the case of ^{199}Hg, is most sensitive to nuclear T violation effects. Of more recent interest are paramagnetic molecules such as PbF and HgF, and of particular current interest is YbF. The molecular spectroscopy of the latter molecule has now been studied as a prelude to an EDM experiment. Paramagnetic molecules, as in the case of paramagnetic atoms, are particularly sensitive to an electron EDM.

9.2 TlF Beam Experiments

The TlF experiments are technically more complicated than any of the others discussed in this book. Nonetheless, the general features are simple, and a qualitative explanation of the operation is possible.

Table 9.1 summarizes the TlF beam experiments. For these experiments, the initial beam state selection was done electrostatically; the two earliest experiments, done at Oxford, used deflection in an electric field gradient, while the later experiments employed quadrupole focussing. This was a marked improvement in the later experiments, and we will concentrate on their description, other than to mention that the Oxford experiment of 1980 had an interaction region 10 m long. Unfortunately, this experiment did not give the anticipated increase in sensitivity because of a low rate of detection and a high background.

The experimental goal is to measure the energy associated with changing the direction of the Tl nuclear spin $\boldsymbol{I}_1$ relative to the intranuclear axis, $\boldsymbol{\lambda}$, as described by the Hamiltonian

$$H = \mu_1 \boldsymbol{\sigma}_1 \cdot \boldsymbol{B}_0 - \nu(\mathrm{Tl}) \boldsymbol{\sigma}_1 \cdot \boldsymbol{\lambda}. \tag{9.3}$$

The first term is the usual hyperfine interaction of the Tl nuclear moment μ_1 with the internal molecular magnetic field $\boldsymbol{B}_0$, where $\boldsymbol{\sigma}_1$ are the Pauli matrices and describe the spin-1/2 Tl nucleus. The second term describes a possible P and T violating interaction; $\boldsymbol{\lambda}$ is the unit vector of the intranuclear axis, and $\nu(\mathrm{Tl})$ is the strength of the possible new effect. In free space, this interaction would produce a small permanent molecular EDM by tipping the intermolecular axis, giving it a small projection along $\boldsymbol{\sigma}_1$. However, it is unlikely that such an effect could be measured with any appreciable accuracy. A cleaner approach is to look for a correlation of the nuclear spin flip frequency as a function of the intranuclear axis orientation relative to $\boldsymbol{B}_0$.

The important point is that the internal energy states should be independent of the orientation of the intranuclear axis relative to the direction of the angular momentum of the molecule. Any such effect would be direct evidence for P and T violation.

Table 9.1. History of TlF experiments

Year	Limit, $\Delta\nu_{\mathrm{exp}}$ 10^{-3} Hz	Reference	Comments
1969	300	[295]	
1980	100	[296]	Multichannel source, 10 m E-field region
1983	30	[297]	Focussing polarizer, $\Delta E < 2$ V/cm
1987	6	[298]	Focussing polarizer
1989	0.6	[299]	Focussing polarizer, jet source, $\Delta E < 0.2$ V/cm, magnetization flip

Before delving into the details of the experiment, consider first the spin–rotational structure of the TlF $^1\Sigma$ electronic ground state [300, 301]:

$$H_{\mathrm{sr}} = H_{\mathrm{rot}} + H_{\mathrm{St}} + H_{\mathrm{Z}} + H_{\mathrm{sr}} + H_{\mathrm{ss}}, \tag{9.4}$$

where the terms are as follows:

1. Rotational energy

$$H_{\mathrm{rot}} = BL(L+1).$$

2. The molecular Stark interaction for an externally applied field $\boldsymbol{E}$ acting on the molecular electric dipole moment d

$$H_{\mathrm{St}} = d\,\boldsymbol{\lambda}\cdot\boldsymbol{E}.$$

3. The Zeeman interaction with an external field $\boldsymbol{B}$ (1 and 2 refer to the Tl and F nuclei, respectively, and $\boldsymbol{\mu}_1 = \mu_1 \boldsymbol{I}_1 \ldots$)

$$H_{\mathrm{Z}} = -\left(\frac{\boldsymbol{\mu}_L}{L} + \frac{\boldsymbol{\mu}_1}{I_1} + \frac{\boldsymbol{\mu}_2}{I_2}\right)\cdot\boldsymbol{B}.$$

4. The interaction of the nuclear spin with the molecular rotational magnetic field

$$H_{\mathrm{sr}} = (c_1 \boldsymbol{I}_1 + c_2 \boldsymbol{I}_2) \cdot \boldsymbol{L}.$$

5. The spin–spin interaction

$$H_{\mathrm{ss}} = 5c_3 \left[\frac{3(\boldsymbol{I}_1 \cdot \boldsymbol{L})(\boldsymbol{I}_2 \cdot \boldsymbol{L}) + 3(\boldsymbol{I}_2 \cdot \boldsymbol{L})(\boldsymbol{I}_1 \cdot \boldsymbol{L}) - 2(\boldsymbol{I}_1 \boldsymbol{I}_2) L(L+1)}{(2L+3)(2L-1)} \right]$$
$$+ c_4 \boldsymbol{I}_1 \cdot \boldsymbol{I}_2.$$

The parameters for this Hamiltonian are [300, 301]

$B = 6.68992$ GHz
$\mu_L = 35(15)$ Hz/G (205)
$\mu_1 = 1.2405(3)$ kHz/G (203)
$\mu_1 = 1.2285(3)$ kHz/G (205)
$\mu_2 = 2.00363(4)$ kHz/G

$d = 4.2282(8)$ Debye
$c_1 = 126.03(12)$ kHz
$c_2 = 17.89$ kHz
$c_3 = 0.70(3)$ kHz
$c_4 =$-13.30(72) kHz.

There is a single F isotope, and natural Tl comprises 70% mass 205 and 30% mass 203.

TlF beams are easily formed by heating a sample to about 650 K; the fractional population in the $L = 1$ rotational state is about 10^{-4}. In the most recent experiment at Yale (Fig. 9.1), this was increased by use of a supersonic source, contributing to a factor 100 increase in beam flux through the apparatus as compared to the 1987 experiment.

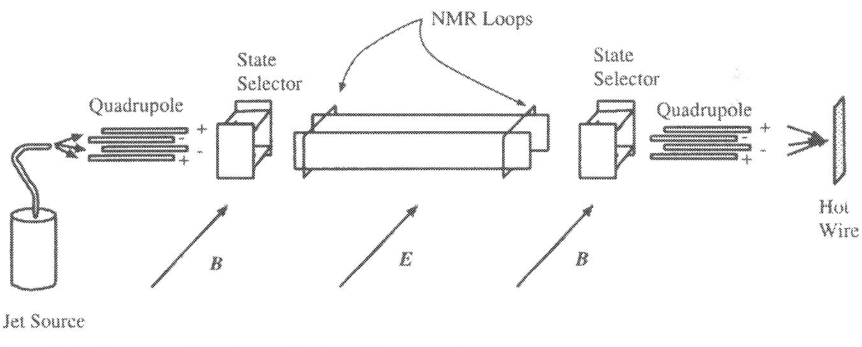

Fig. 9.1. Schematic of the Yale TlF experiment

After the beam leaves the oven slit and collimation, molecules in the $L = 1$, $m_L = 0$ state are focussed by an electrostatic quadrupole field generated by four rods parallel to the beam, with the sign of the applied potentials alternating between each rod, and of length a few tens of cm. States with $m_L \neq 0$ are defocussed, and $L > 1$ states are not effectively focussed. At this point, there is no selection of nuclear spin.

Since the quadrupole state selector cannot distinguish the nuclear spin orientations and only focusses the $L+1$, $m_L = 0$ state, a "triple resonance" scheme must be used. Furthermore, the experimental observable is an interference between a possible EDM precession in the internal molecular electric field, and the usual spin precession in the molecular magnetic field. For convenience, we label the $L = 1$ states as (m_L, m_1, m_2) where $m_1 = 2s_{z1}$, $m_2 = 2s_{z2}$ refer to the Tl and F spin projections. After leaving the quadrupole focusser, molecules enter a region of applied static electric and magnetic, and oscillating electric, fields which can drive transitions between the various hyperfine levels within the $L = 1$ manifold.

In the most recent experiment, the state selector incorporated 7 G static magnetic and 35 V/cm static electric fields, so all of the hyperfine levels were resolved; transitions between them could be driven with an oscillating electric field. Two transitions were used for EDM measurements:

case 1, with $(0, -1, -1) \rightarrow (-1, 1, -1)$;

case 2, with $(0, 1, 1) \rightarrow (1, -1, 1)$.

The electric field between the state selector and the holding field had to be tailored to ensure adiabatic passage between the regions.

After leaving the state selector, the molecules encounter a first RF separated oscillatory field coil and strong electric field region (about 30 kV/cm, resulting in a 70% axis alignment efficiency). The drive frequency and amplitude is set to give a $\pi/4$ pulse for the Tl nucleus in the internal molecular field, about 120 kHz. Unfortunately, due to a polarizability effect, this frequency depends on the amplitude of the applied electric field (41 mHz/(V/cm) at 29.5 kV/cm) and required that this field be reversed to an accuracy of order 0.1 V/cm (however, comparison between case 1 and case 2 could discriminate this effect).

After passing through the interaction region of length approximately 2.5 m, giving a coherence time of about 6 ms for the 400 m/s supersonic source TlF molecules (linewidth of about 100 Hz), a second separated oscillatory field coil was encountered. If the Tl nuclear spin did not return to its original state, on encountering a second identical state selector, a transition to the original $m_L = 0$ state would not occur, and the molecule would not be focussed onto the detector by a second electrostatic quadrupole. If the molecules returned to their original state after the second $\pi/4$ pulse, they would be focussed onto the detector.

The typical detected flux was about 4×10^7/s; a hot tungsten wire, oxygen activated, gave nearly 90% detection efficiency, and the background rate was about twice the TlF flux. The hot wire detector was operated in the current mode; TlF dissociating on the wire produced a Tl^+ ion, leaving an electron behind in the wire; this current was measured with an electrometer.

Systematic effects were studied with great care and a number of reversals were used to detect and discriminate them from a true T violation. These reversals included measurements with the case 1 and case 2 states corresponding to an internal molecular magnetic field reversal, holding electric field reversal, reversing all the angular momenta for the various states, and reversing the polarity of the quadrupole field. Leakage current magnetic fields could couple through the H_Z term in the Hamiltonian, but were insignificant as determined from the known magnitude of these currents. A most interesting point is that this experiment was essentially free from any motional field effect. The total magnetic field affecting the Tl nucleus is given by the sum of the small external field (typically 1 mG) and the relatively large internal molecular field (about 50 G) which is identically parallel to the applied electric field. Thus, the net magnetic field is perpendicular to the motional field to an accuracy of $10^{-3}/50 \approx 10^{-5}$ rad. Furthermore, the applied electric field was relatively weak, resulting in a motional field magnitude of 0.155 mG. This results in a quadratic shift of the Tl resonant frequency of 0.2 μHz, which was unobservable in this case. The final result was reported as [299]

$$\nu = (-0.13 \pm 0.22)\ \text{mHz} \rightarrow \nu < 0.4\ \text{mHz} \tag{9.5}$$

for the Tl^+ nuclear spin flip parameter given in (9.3).

This type of experiment could be improved by at least an order of magnitude before systematic effects become uncontrollable. A group at the Petersburg Nuclear Physics Institute has proposed a carrier gas jet supersonic TlF source that might give two orders of magnitude higher beam intensity [302]. Also proposed is a zero-magnetic-field interference experiment in the $m_L = 0$ hyperfine states; this eliminates the Stark effect described above. This experimental technique has now been demonstrated [307].

With a slower beam and increased flux, along with improved focussing in the apparatus, it has been estimated that a three order of magnitude improvement in statistical noise, to the level of 10^{-7} Hz, might be possible. Of course, in this case, the accuracy of the experiment may be limited by systematic effects.

9.3 What Have We Learned from the TlF Experiment?

A serious problem with the TlF experiments is that up to now there is no reliable way to obtain from them quantitative limits on CP violation. In this respect the molecular experiments differ essentially from atomic ones where in most cases such limits have been derived. Elaborate calculations of the T-odd effects in the TlF molecule have been performed [272, 291, 303, 304]. However, the results of even the most detailed of these calculations, in the authors' opinion, "must be regarded as slightly greater than order-of-magnitude estimates" [304]. In order to interpret the experimental results at least qualitatively, we will resort in this section to a rather crude model of the TlF molecule, presenting afterwards the predictions of [304].

The model is as follows. We will assume that in the TlF molecule the external $6p$ electron of the thallium atom is bound to fluorine. As a result the Tl^+ ion is in the electric field $E = -e/r_1^2$, which polarizes its external $6s^2$ shell. According to [305], in this molecule the internuclear distance is $r_1 = 3.92\, a$. Therefore, the one-particle wave functions of the $6s$ electrons in the Tl^+ ion acquire an admixture of the $6p$ states:

$$|\,\Omega\rangle = |\,6s,\ \Omega\rangle + \beta\left(-\frac{2\Omega}{\sqrt{3}}\,|\,6p_{1/2},\ \Omega\rangle + \sqrt{\frac{2}{3}}\,|\,6p_{3/2},\ \Omega\rangle\right); \tag{9.6}$$

$$\beta = \frac{2}{\sqrt{3}}\,\frac{\mathrm{Ry}}{E_{6s} - E_{6p}}\,\frac{a^2\rho(6s,6p)}{r_1^2} = 0.27.$$

Here $\Omega = \pm 1/2$ is the projection of the one-electron angular momentum on the molecular axis; the radial integral $\rho(6s,6p) \approx -2.3$; the one-electron energies of the ion have the form $E = -Z_i^2\mathrm{Ry}/\nu^2$ where the effective charge of the Tl^+ ion $Z_i = 2,\ \ \nu_{6s} = 1.63,\ \ \nu_{6p} = 2.15$.

It is convenient to start from the interpretation of the experimental results in terms of the CP-odd tensor electron–nucleon interaction. According to the simple shell model, the state of a valence proton in the thallium nucleus is $s_{1/2}$, so that the nuclear matrix element

$$\left\langle k_{2p}\sum_p \boldsymbol{\sigma}_p + k_{2n}\sum_n \boldsymbol{\sigma}_n\right\rangle$$

reduces to $k_{2p}\boldsymbol{I}/I$. This value can be refined using the experimental value of the magnetic moment of the nucleus, as has been done in the previous chapter for ^{129}Xe. As a result, the effective constant multiplying $\boldsymbol{I}/I$ is equal to

$$k_2(\mathrm{Tl}) = 0.75\,k_{2p} + 0.25\,k_{2n}.$$

Using the tensor part of the matrix element (8.25) and employing (9.6), one can easily find the following expression for the effective Hamiltonian of the CP-odd interaction of the thallium nucleus with the molecular axis $\boldsymbol{\lambda}$:

$$H_{eff} = \nu(\mathrm{Tl})\,\boldsymbol{\lambda}\cdot\boldsymbol{I}/I;$$

$$\nu_k(\mathrm{Tl}) = \frac{Gm^2\alpha^2}{\sqrt{2}\pi}\,\frac{Z^2R}{(\nu_s\nu_p)^{3/2}}\,\frac{8\beta}{\sqrt{3}}\,Z_i^2\,\mathrm{Ry}\,k_2(\mathrm{Tl}) \tag{9.7}$$

$$= 5\times 10^3\,k_2(\mathrm{Tl})\ \mathrm{Hz}. \tag{9.8}$$

Since this is clearly only a rough estimate, to quantitatively interpret the experimental data, we will use the result of the most detailed calculations [304, 306, 307] which gives for the quantity $\nu_k(\mathrm{Tl})$ a value almost 6 times smaller. With this correction factor, the experimental number (9.5) leads to the following upper limit:

$$k_2(\mathrm{Tl}) < 4.5\times 10^{-7}. \tag{9.9}$$

Using now relations (8.46), (8.53) and (8.69) from the previous chapter, we also obtain upper limits on other CP-odd parameters:

$$k_3(\mathrm{Tl}) = 0.75\,k_{3p} + 0.25\,k_{3n} < 3 \times 10^{-4}; \tag{9.10}$$

$$k_1(\mathrm{Tl}) = 0.4\,k_{1p} + 0.6\,k_{1n} < 1.7 \times 10^{-5}; \tag{9.11}$$

$$d_e < 10^{-24}\,e\,\mathrm{cm}. \tag{9.12}$$

And finally, let us obtain from the experimental limit (9.5) the constraint on the Schiff moment. The SM contribution to the effective constant ν, as derived with formulas (8.77), (9.6), is

$$\nu_S(\mathrm{Tl}) = \frac{S(\mathrm{Tl})}{\pi e a^3}\,\frac{Z^2 Z_i^2 (R_{1/2} + 2R_{3/2})}{(\nu_s \nu_p)^{3/2}}\,\frac{8\beta}{3\sqrt{3}}\,Z_i^2\,k_2(\mathrm{Tl}) \tag{9.13}$$

$$= 1.6 \times 10^5\ \mathrm{Hz} \times \frac{S(\mathrm{Tl})}{e}\ \mathrm{fm}^3. \tag{9.14}$$

In this case the detailed calculation [304] gives a $\nu_S(\mathrm{Tl})$ value which is nearly 3 times smaller. Allowing for this correction factor 1/3, we obtain from the experimental limit (9.5)

$$S(\mathrm{Tl}) < 0.8 \times 10^{-8}\,e\,\mathrm{fm}^3. \tag{9.15}$$

Since

$$S(\mathrm{Tl}) = -\frac{2\pi}{3}\,\delta r^2\,d_\mathrm{p},$$

assuming for $\delta r^2 = r_q^2 - r_d^2$ the value 4 fm^2, obtained with the Woods–Saxon potential, we arrive at the following upper limit on the proton EDM:

$$d_\mathrm{p} < 10^{-22}\,e\,\mathrm{cm}. \tag{9.16}$$

Even allowing for all the uncertainties mentioned, it is the best upper limit on the proton dipole moment.

9.4 Paramagnetic Molecules

If there is an unpaired electron in a polar molecule, a similar type of experiment can be extraordinarily sensitive to the electron EDM [270]. In this case, the usual relativistic enhancement occurs, in addition to the effective enhancement of the applied laboratory field. Taking these two factors together, for the case of HgF, the net gain is 6×10^7! Furthermore, this type of experiment is relatively immune to the motional field effects that complicate the Tl beam electron EDM experiments. Comparing HgF to Tl, the relativistic enhancement factors are about equal, but the internal molecular electric field (established by a weak external field) and the applied electric field in the Tl EDM experiment differ by a factor

$$\frac{10^9\mathrm{V/cm}}{100\mathrm{kV/cm}} = 10^4 .$$

It is reasonable to assume the same coherence time, and similar polarization/analysis efficiency. The counting rate in the Tl EDM experiment was about 10^8/s, and the figure of merit for a given experiment is given roughly by $E\sqrt{I}$, as given by (3.6). Thus, the current sensitivity to an electron EDM could be obtained with a factor of 10^8 lower detected flux, or, in comparison to the counting rate on the Tl EDM experiment, 1 count/s.

Unfortunately, paramagnetic molecules are not only very reactive, but also thermodynamically unstable, and there is scant information about their characteristics. Recently, a beam of YbF molecules was produced with a view to the development of an electron EDM experiment [308]. YbF, although suffering a factor of three lower enhancement compared to HgF [309], has the advantage that it can be coaxed into forming a molecular beam. Attempts to form a HgF beam were not successful, primarily because its partial pressure in equilibrium with Hg, F_2, F, and HgF_2 is exceedingly low [307] simply because HgF_2 is the more stable state.

A YbF beam can be produced by heating YbF_3 with Al to 1200 °C in a sealed stainless steel tube; a 1 mm hole in an endcap of the tube defines the molecular beam. Laser doppler–free spectroscopy was used to investigate the 0–0 band of the $A^2\Pi_{1/2} - X^2\Sigma^+$ transition, around 552 nm. Some very interesting features in the Σ^+ state were found [308]: The spin–rotation coupling constant is anomalously small, about 13 MHz. The centrifugal distortion factor, describing the change in the spin–rotation coupling with increasing rotational quantum number, is negative. This is the only known example of such behavior. Clearly, the YbF molecule has interest beyond EDM work. Hopefully, an experimental EDM technique based on laser spectroscopy will be developed to take advantage of the enormous enhancement that this amazing system offers.

9.5 What Will Be Gained from Experiments with Paramagnetic Molecules?

Let us consider the above-mentioned enhancement in more detail. The electron ground state of YbF is $^2\Sigma^+$ (the superscript $^+$ here means that the wave function does not change sign on reflection in a plane passing through the axis of the molecule). We will assume the following model of the molecule. One of the electrons of the Yb atom (its ground state configuration is $4f^{14}\,6s^2$) leaves it for the fluorine which thus becomes a negative ion with a closed p shell. Then the electron coordinate wave function can be expanded in the wave functions of the Yb^+ ion:

$$|\Sigma\rangle = x_s|s\rangle + x_p|p\rangle + x_d|d\rangle + x_f|f\rangle + \dots . \tag{9.17}$$

Since we are dealing with a Σ term, in the expansion (9.17) even the terms with nonzero orbital angular momentum, $|p\rangle$, $|d\rangle$, etc, have $l_z = 0$.

The interaction of the spin with the rotation of the molecule in a $|\Sigma\rangle$ state is small ([15], §84). It is controlled in the $^2\Sigma$ fluorides by the combined effect of the Coriolis interaction, which is $-2B\boldsymbol{J} \cdot \boldsymbol{l}$ as long as the electron spin is decoupled, and of the spin–orbit interaction [308, 310, 311]. As mentioned in the previous section, the strength of this interaction in YbF is anomalously small, much less than in lighter $^2\Sigma$ fluorides. This seems mysterious since the spin–orbit interaction grows with Z. The natural clue to the mystery is that, as distinct from lighter metals forming $^2\Sigma$ fluorides, ytterbium has the inner f shell, and moreover, this shell can be easily excited. Indeed, the first excitation of the Yb^+ ion (its ground state is $4f^{14}\,6s$) is the f hole state, $4f^{13}\,6s^2$, $J = 7/2$, which even lies below the simplest excitation, $4f^{14}\,6p$ [312]. The configuration $4f^{13}\,6s^2$ has a negative fine-structure splitting, as expected for a hole state. Not only is this splitting large, but it enters the discussed spin–rotational constant multiplied by $l(l+1) = 12$, which is much larger than the corresponding values of $l(l+1)$ for lower l. Therefore, an admixture of the Yb^+ states with a $4f$ hole in the electron $|\Sigma\rangle$ state [the last term of the expansion (9.17)] explains in a natural way the discussed anomaly [313]. Even entering with a relatively small weight, this admixture can strongly compensate the positive contributions of usual excitations, $4f^{14}\,6p$, $4f^{14}\,5d$, etc, (second and third terms in the expansion (9.17)) to the spin–rotational constant. The same admixture can also explain the negative centrifugal distortion factor in the spin–rotation coupling. One more argument in favor of this explanation of both anomalies is the recent discovery of an $\Omega = 1/2$ state nearly degenerate with $A\,^2\Pi_{1/2}$, whose vibrational and rotational frequencies are consistent with those of an $f^{13}\,s^2$ configuration [314].

But let us return to CP violation and to the EDM induced by the electron dipole moment in the ground state of the YbF molecule. As in our estimates for diamagnetic TlF, we will restrict ourselves to $6s$ and $6p$ orbitals in the expansion (9.17) and present the truncated wave function (including spin) as

$$x_s|\,6s,\ \Omega\rangle + x_p\left(-\frac{2\Omega}{\sqrt{3}}\,|\,6p_{1/2},\ \Omega\rangle + \sqrt{\frac{2}{3}}\,|\,6p_{3/2},\ \Omega\rangle\right). \tag{9.18}$$

But as distinct from TlF, in this case the values of coefficients $x_{s,p}$ will be taken from [315] (where they were calculated in fact for generalized s, p orbitals):

$$|x_s|^2 = 0.84, \quad |x_p|^2 = 0.15.$$

Using the formula (8.13), we now obtain the following expression for the energy shift due to the electron EDM:

$$\delta E = -(\boldsymbol{\sigma}\cdot\boldsymbol{\lambda})\,\frac{16}{\sqrt{3}}\,x_s x_p\,\frac{Z^3\alpha^2 Z_i^2}{\gamma(4\gamma^2-1)(\nu_s\nu_p)^{3/2}}\,\frac{d_e}{ea}\,\mathrm{Ry}. \tag{9.19}$$

As in the TlF case, the energy of the Yb^+ ion is presented as $E = -Z_i^2\mathrm{Ry}/\nu^2$; $Z_i = 2, \;\; \nu_{6s} = 1.63, \;\; \nu_{6p} = 2.15$.

The effective electric field acting on the electron EDM, estimated in this way, is

$$\frac{\delta E}{d_e} \sim 3 \times 10^{10} \, \frac{\mathrm{V}}{\mathrm{cm}}. \tag{9.20}$$

The gain in the field strength, as compared to the 5 kV/cm needed to achieve the YbF alignment close to unity, is roughly 6×10^6!

If we take for the value of the electric EDM the best present limit (6.5), $d_e < 4 \times 10^{-27}\, e\,\mathrm{cm}$, the frequency shift, corresponding to (9.20), equals 30 mHz.

More elaborate calculations of the electron EDM enhancement in YbF are presented in [309]. The coefficients of the molecular wave function expansion in atomic orbitals (which does not include f hole states) are estimated there by fitting the YbF hyperfine structure. Our estimate (9.20) is in excellent agreement with their result.

In the same way one can estimate the YbF EDM as induced by the contact CP-odd electron–nucleon interaction (see also [309]).

YbF molecules with odd ytterbium isotopes (^{171}Yb, $I = 1/2$ and ^{173}Yb, $I = 5/2$) may turn out to be of special interest. The nonzero nuclear spin of these nuclei would allow the investigation of CP violation in the hadronic sector. The corresponding effects in paramagnetic molecules are discussed in numerous papers [278,316–320]. Although in ^{171}Yb one can count only on the nuclear Schiff moment, in ^{173}Yb the magnetic quadrupole moment becomes operative, bringing an extra order of magnitude gain in the induced EDM of the molecule. Unfortunately, the natural abundance of the odd ytterbium isotopes is not high: 14% for ^{171}Yb, and 16% for ^{173}Yb. These experiments demand resolution of the molecular hyperfine structure. In the case of ^{173}Yb this results in another serious loss in statistical accuracy due to the large number of hyperfine components, along with the loss due to the usual finite-temperature rotational state distribution.

10. CP-Odd Nuclear Forces

This is one more question: where do answers come from?

A. Strugatsky and B. Strugatsky

The upper limit (8.93) for the neutron EDM d_n derived from atomic experiments is by an order of magnitude worse than the result (4.43) of the direct measurement. For the proton dipole moment d_p, the molecular TlF result (9.16) is even more modest. Of course, formally speaking, this is by itself the best experimental upper limit on the proton EDM. However, one can hardly imagine a situation where the proton and neutron dipole moments could differ by several orders of magnitude, since proton and neutron are strongly interacting particles which can easily transform virtually into one another (for instance, $n \rightleftharpoons p\pi^-$, $p \rightleftharpoons n\pi^+$). For the electron EDM, at least in the standard model, the theoretical prediction is many orders of magnitude smaller than that for the neutron dipole moment (although in many speculative extensions of the standard model, this in not the case). And finally, the possible values of the CP-odd electron–nucleon interaction constants $k_{1,2,3}$ in most theoretical models lie far away from the experimental limits. Therefore, the impression may arise that searches for T-odd effects in atoms and molecules are of no serious interest at all for elementary particle physics.

However, the real state of affairs is quite different. It has been demonstrated [278] that the nuclear dipole moments induced by the T- and P-odd nuclear forces can be about 100 times larger than the nucleon EDM. Such a relation between the dipole moments of nuclei and an individual nucleon, sharply increases the significance of the atomic and molecular experiments for elucidating the origin of CP-violating interactions; even in the hadronic sector they provide comparable information to the neutron experiments.

10.1 CP-Odd Mixing of Opposite-Parity Nuclear Levels

Let us start by discussing the phenomenological T- and P-odd nucleon–nucleon potential. If we assume for simplicity the interaction to be local and limit ourselves to first-order terms in the nucleon velocities p/m_p, then to this approximation the most general form of the effective potential (in the spirit of the Landau–Migdal approach [321]) is

$$W_{ab} = \frac{G}{\sqrt{2}}\frac{1}{2m}\{(\xi_{ab}\boldsymbol{\sigma}_a - \xi_{ba}\boldsymbol{\sigma}_b)\boldsymbol{\nabla}\,\delta(\boldsymbol{r}_a - \boldsymbol{r}_b)$$

$$+\xi'_{ab}(\boldsymbol{\sigma}_a \times \boldsymbol{\sigma}_b)[(\boldsymbol{p}_a - \boldsymbol{p}_b)\,\delta(\boldsymbol{r}_a - \boldsymbol{r}_b) + \delta(\boldsymbol{r}_a - \boldsymbol{r}_b)\,(\boldsymbol{p}_a - \boldsymbol{p}_b)]\}\,. \quad (10.1)$$

The dimensionless constants ξ, characterizing the strength of the interaction in units of the Fermi constant G, are supplied with subscripts in order to distinguish between protons and neutrons. These are effective constants and already include the exchange terms for identical nucleons. In a detailed theory the constants should also include additional suppression factors reflecting the long-range and exchange nature of the realistic interaction, as well as the short-range nucleon–nucleon repulsion. As must be the case, the potential (10.1) is invariant under the Galilean transformations. In fact, expression (10.1) is nothing else but the nonrelativistic limit of the most general form of the T- and P-odd four-fermion interaction considered in Sect. 2.3. This is obtained when the general interaction is restricted to terms linear in the momenta of both particles, and written in a coordinate representation.

Let us now restrict ourselves to the simplest approximation of the shell model, assuming that the nuclear spin coincides with the total angular momentum of an odd valence nucleon, while the other nucleons form a core with the zero angular momentum. If we consider also only those contributions to the CP-odd nuclear multipoles which are induced by the admixture of the valence nucleon states without the core excitation, then the T- and P-odd interaction reduces to

$$W_1 = \frac{G}{\sqrt{2}}\frac{1}{2m}\sum_b \xi_{ab}\boldsymbol{\sigma}_a\boldsymbol{\nabla}\,\delta(\boldsymbol{r}_a - \boldsymbol{r}_b). \quad (10.2)$$

To obtain expression (10.2) we have identified in formula (10.1) the nucleon a with the valence nucleon and summed over all the nucleons b of the core, omitting the terms dependent on $\boldsymbol{p}_b$ and $\boldsymbol{\sigma}_b$. Now, averaging expression (10.2) over the state of the core nucleons, we arrive at the following effective T- and P-odd single-particle potential for the outer nucleon:

$$W = \frac{G}{\sqrt{2}}\frac{\xi_a}{2m}\,\boldsymbol{\sigma}\,\boldsymbol{\nabla}\rho(r) = \frac{G}{\sqrt{2}}\frac{\xi_a}{2m}\,\mathrm{i}\,[\boldsymbol{\sigma p}, \rho(r)]. \quad (10.3)$$

Here $\rho(r)$ is the density of nucleons in the core normalized by the condition $\int \mathrm{d}\boldsymbol{r}\rho(r) = A$ (the atomic number is assumed to be large, $A \gg 1$);

$$\xi_a = \xi_{ap}\frac{Z}{A} + \xi_{an}\frac{N}{A},$$

the subscript a on ξ_a takes the values p and n for an outer proton and neutron respectively; on the spin, coordinate and momentum of the valence nucleon this subscript is omitted; [,] denotes a commutator.

In heavy nuclei the CP-odd mixing matrix elements as induced by the operators (10.1) and (10.3) is parametrically suppressed as $A^{-1/3}$ [322–325]. Indeed, a heavy nucleus is a semiclassical system, the corresponding large parameter being $A^{1/3}$. Since a commutator has no classical limit, it is only natural to expect that the matrix element of operator (10.3) is suppressed just by this parameter. The same conclusion can be anticipated due to the presence of $\nabla\rho(r)$ in operator (10.3); the mixing is a surface, not a volume, effect.

Let us pass on now from general arguments to a simple estimate. Assuming for the core density a step-like profile

$$\rho(r) = \rho_0\,\theta(R-r), \tag{10.4}$$

we get

$$\langle 2|\nabla\rho(r)|1\rangle \approx \rho_0\,\Re_2\Re_1 R^2. \tag{10.5}$$

Here $R = r_0\,A^{1/3}$ is the nuclear radius ($r_0 = 1.2$ fm), $\Re_{2,1}$ are the values of the radial wave functions of a valence nucleon in the states $|2\rangle$, $|1\rangle$ at $r = R$. The following estimate is well known (see [326], Vol. 1, Chap. 2)

$$\Re^2 R^3 \approx 1.5$$

(which means that the nucleon density at the boundary constitutes one half of the internal density). In this way we get

$$\langle 2|W|1\rangle \approx \frac{G}{\sqrt{2}}\,\frac{\xi_a}{2m}\,\rho_0\,r_0^{-1}1.5\,A^{-1/3}. \tag{10.6}$$

Numerically,

$$\langle W\rangle \approx 1\text{ eV}\,\xi_a A^{-1/3}. \tag{10.7}$$

A simple closed form for the CP-odd admixture to a nuclear wave function, caused by the interaction (10.3), can be derived in a model approach [278], which, though being approximate, allows one to obtain analytical results for CP-odd effects in nuclei and investigate their parametrical dependence. Let us suppose that the outer nucleon moves in a spherically symmetric potential $U(r)$. Since the profiles of the nuclear density $\rho(r)$ and potential $U(r)$ are similar, let us assume that they coincide exactly:

$$\rho(r) = U(r)\,\frac{\rho_0}{U_0}.$$

Then the potential (10.3) can be rewritten as

$$W(\boldsymbol{r}) = \lambda\,\boldsymbol{\sigma}\cdot\nabla U(r), \quad \lambda = \frac{G}{\sqrt{2}}\,\frac{\xi}{2m}\,\frac{\rho_0}{U_0} = -\,2\times 10^{-21}\,\xi\text{ cm}. \tag{10.8}$$

Accordingly, the total potential in which the nucleon moves is

$$\tilde{U}(\boldsymbol{r}) = U(r) + W(\boldsymbol{r}) = U(r) + \lambda\,\boldsymbol{\sigma}\cdot\nabla U(r), = U(|\boldsymbol{r} + \lambda\,\boldsymbol{\sigma}|). \tag{10.9}$$

In this potential, it is obvious that the wave function of the external nucleon becomes

$$\tilde{\psi}(\boldsymbol{r}) = \psi(\boldsymbol{r} + \lambda \boldsymbol{\sigma}) = (1 + \lambda \boldsymbol{\sigma} \cdot \boldsymbol{\nabla})\, \psi(\boldsymbol{r}), \tag{10.10}$$

where $\psi(\boldsymbol{r})$ is its unperturbed value.

It can be easily demonstrated that to reconcile the obtained CP-odd admixture to the wave function with the estimate (10.6), the matrix element (10.6) should not be divided by the typical energy interval ΔE, as is usually the case in a perturbative approach, but by the typical potential depth U_0. This additional suppression $\sim A^{-1/3}$ of the CP-odd mixing corresponds in the perturbative treatment to a strong cancelation between contributions of different admixed states. The cancelation was pointed out long ago [326a].

10.2 Nuclear Moments Induced by *T*- and *P*-Odd Potentials

We will start this section (based essentially on the results obtained in [278]) with the general formulas for the T-odd nuclear moments. Let us present the wave function of the outer nucleon perturbed by the potential (10.3) in the usual form

$$\psi(\boldsymbol{r}) = \left[R_0(r) - \sum_n \eta_n R_n(r)\, (\boldsymbol{\sigma} \cdot \boldsymbol{n}) \right] \Omega_{Ilm}. \tag{10.11}$$

In calculating the nuclear EDM we should bear in mind that since the center of mass of the nucleus stays at rest, a shift of the external nucleon by a distance $\boldsymbol{r}$ results in the shift of the core by $-\boldsymbol{r}/A$. Therefore, the dipole moment operator is (for details see, e.g., [15], §118):

$$\hat{\boldsymbol{d}} = e\,(q - Z/A)\, \boldsymbol{r}, \tag{10.12}$$

where $q = 1$ or 0 for an outer proton or neutron, respectively. Using (10.11) we easily obtain the following expression for the nuclear EDM:

$$d = -e\,(q - Z/A)\, \langle r \rangle\, \frac{1}{2(I+1)}. \tag{10.13}$$

Hereafter, we write

$$\langle r^k \rangle = \sum_n \eta_n \int_0^\infty \mathrm{d}r\, r^{k+2} R_0(r)\, R_n(r). \tag{10.14}$$

The nuclear SM can be written through considerations similar to those used in Chap. 8 as:

$$S = -\frac{\pi\, e\, q}{I+1} \left(\frac{1}{5} \langle r^3 \rangle - \frac{1}{3} r_q^2 \langle r \rangle \right). \tag{10.15}$$

Curiously, in the case of a SM in the approximation used here, the recoil effects for the first and second terms in (10.15) cancel. This can be easily shown by direct calculations.

And finally, for calculating the nuclear MQM, as defined by the formula

$$M_{mn} = \int d\boldsymbol{r}\,(r_m \epsilon_{nrs} + r_j \epsilon_{irs})\, r_r j_s, \tag{10.16}$$

we use the expression

$$\boldsymbol{j} = \frac{eq}{2m}[\psi^\dagger(-\mathrm{i}\boldsymbol{\nabla})\,\psi + (-\mathrm{i}\boldsymbol{\nabla}\psi)^\dagger\,\psi] + \frac{e\mu}{2m}\,\boldsymbol{\nabla}\times\psi^\dagger\boldsymbol{\sigma}\psi$$

for the current density of the outer nucleon, where μ is its total magnetic moment. As a result, we get

$$M_{mn} = \frac{e}{2m}\left\langle 3\mu\left[r_m\sigma_n + r_n\sigma_m - \frac{2}{3}\delta_{mn}(\boldsymbol{\sigma r})\right]\right.$$
$$\left. + 2q\,(r_m l_n + r_n l_m)\right\rangle; \tag{10.17}$$

$$\mathcal{M} = \frac{e}{2m}\,(\mu - q)\,\langle r\rangle\,\frac{2I-1}{I+1}. \tag{10.18}$$

In formula (10.17), l_n is the orbital angular momentum of the outer nucleon. Let us note that according to (10.18) the MQM of a spherical nuclei in the adopted single-particle description is proportional to the anomalous magnetic moment of the external nucleon. This statement can be easily checked by means of the Dirac equation in a spherical potential. But in a nonspherical nucleus, the normal magnetic moment also contributes to the MQM.

One can, of course,. determine the mixing coefficients η_n, and then $\langle r\rangle$, $\langle r^3\rangle$, and the CP-odd nuclear moments numerically, using, for example, the Woods–Saxon potential for an external nucleon. However, after the fairly crude approximations with which we managed to go from (10.1) to the single-particle potential (10.3), it is only natural at least to start with closed analytical expressions for the CP-odd nuclear moments obtained by means of the model wave function (10.10). It turns out that the difference between these results and those calculated numerically does not exceed 50%.

In this way we obtain

$$d = -e\,(q - Z/A)\,\lambda\,\frac{1/2-K}{I+1}, \tag{10.19}$$

$$S = -2\pi e q\lambda\left[\frac{1}{3}\left(K - \frac{1}{2}\right)r_q^2 - \frac{1}{5}\left(K - \frac{3}{2}\right)r_d^2\right]\frac{1}{I+1}, \tag{10.20}$$

$$\mathcal{M} = \frac{e}{m}\,(\mu - q)\,\lambda\,\frac{(2I-1)(1/2-K)}{I+1}. \tag{10.21}$$

Bearing in mind that

$$r_d^2 \approx r_q^2 \approx \frac{3}{5}\,r_0^2\,A^{2/3},$$

expression (10.20) for the SM can be rewritten in a more compact form:

$$S = -\frac{4\pi}{25}\, e\, q\, \lambda\, r_0^2\, A^{2/3}\, \frac{K+1}{I+1}. \tag{10.22}$$

The characteristic numerical values of the nuclear T- and P-odd multipoles have the following orders of magnitudes:

$$d/e \approx 10^{-8}\ \mathrm{fm}\ \xi, \tag{10.23}$$

$$S/e \approx 10^{-8}\ \mathrm{fm}^3\ A^{2/3}\xi \approx 10^{-7}\ \mathrm{fm}^3\ \xi, \tag{10.24}$$

$$\mathcal{M}/e \approx 10^{-8}\ \mathrm{fm}^2\ \xi. \tag{10.25}$$

Estimates (10.23-25) can be obtained directly by dimensional arguments. The Fermi (fm) is a natural unit of length for a nucleus, taken to the power which corresponds to the usual dimension of a particular multipole. To explain the factor 10^{-8}, let us recall that $Gm_\pi^2 \sim 10^{-7}$ is the natural dimensionless measure for the nuclear scale of the usual CP-even weak interaction. However, as pointed out in Section 10.1, in the present CP-odd case we have an additional suppression $\sim \Delta E/U_0$, which is close numerically to an order of magnitude. Even the A dependence can be easily restored: the $A^{-1/3}$ suppression of the CP-odd mixing matrix element indicated in the previous section is compensated for by $A^{1/3}$ from r in the EDM and MQM operators, and overcompensated by A from r^3 in the SM.

Let us mention here the detailed numerical calculations [327] of the MQM induced by the CP-odd interaction (10.3). They were performed in the single-particle approximation and demonstrated that the above model result (10.21) is sufficiently close to that obtained in a Woods–Saxon potential with spin-orbit interaction, being smaller by only $15-20\%$. It is curious that the results obtained with an oscillator potential deviate more from the Woods–Saxon results than formula (10.21). However, many-body corrections to the nuclear MQM were found to be large [327a].

The analytic expression (10.19) is useful, in particular, for a parametric comparison of the neutron EDM d_n with that of a nucleus d. Generally speaking, such a comparison is possible only in the framework of a definite scheme of CP violation. Nevertheless, there is a fairly regular enhancement of d as compared to d_n:

$$3\pi\,(m_\pi^2\, U_0\, r_0^3)^{-1} \approx 60. \tag{10.26}$$

The factor $(m_\pi^2\, U_0\, r_0^3)^{-1}$ here is of a quite obvious nuclear nature, while the geometrical factor 3π arises since d usually originates at the tree approximation, while d_n demands at least a one-loop approximation. It should be emphasized that this enhancement of the CP-odd effects in a nucleus compensates to a great extent for the suppression of the induced atomic EDM caused by the electrostatic shielding. This enhancement is the primary reason why experiments with heavy atoms and molecules can compete with searches for the neutron EDM.

10.3 Enhancement Mechanisms for T- and P-Odd Nuclear Multipoles

In principle, the effects of CP violation in nuclei can be additionally enhanced by an anomalous closeness of an opposite-parity level with the same angular momentum to the ground state of a nucleus [278, 304, 328, 329]. Unfortunately, if we limit ourselves to heavy stable nuclei, the choice is small. Available are ^{161}Dy, where there is a level $5/2^-$ at an energy separation of 25.7 keV from the $5/2^+$ ground state, and ^{237}Np (ground state $5/2^+$, excited state $5/2^-$, $\Delta E = 59.5$ KeV). There are several more nuclei with $\Delta E \approx 100$ keV (^{153}Eu, ^{155}Gd, ^{163}Dy, ^{233}U). One might expect here an enhancement of the CP-odd effects by two to three orders of magnitude as compared with ordinary nuclei, where $\Delta E \approx 10$ MeV. However, a more detailed analysis [278] has shown that such expectations are too optimistic.

Let us discuss the problem in more detail, following [278]. All the nuclei mentioned are deformed (nonspherical). The T-odd moments d_z, S_z, M_{zz} are calculated in the "frozen" (rotating with the nucleus) reference frame, and the z axis is directed along the nuclear axis. The results are converted to the laboratory frame via the formulae

$$d_{\text{lab}} = \frac{J}{J+1} d_z, \quad S_{\text{lab}} = \frac{J}{J+1} S_z, \quad \mathcal{M}_{\text{lab}} = \frac{J}{J+1} \frac{2J-1}{2J+3} M_{zz}.$$

These expressions are written for the case when the projection Ω of the angular momentum onto the axis of the nucleus has its maximum value J, which takes place for the ground state of the rotational band.

The contribution of the close level of opposite parity $|\bar{\Omega}\rangle$ to the T-odd multipole $\mathcal{O}$ (d_z, S_z, M_{zz}) of the ground state $|\Omega\rangle$ is

$$\mathcal{O} = 2 \frac{\langle \Omega | W | \bar{\Omega} \rangle \langle \bar{\Omega} | \hat{O} | \Omega \rangle}{E_\Omega - E_{\bar{\Omega}}}. \tag{10.27}$$

The matrix elements here are calculated in the Nilsson single-particle oscillator model. Unfortunately, the matrix elements of W, d_z and S_z between the close opposite-parity levels are small. Moreover, they cannot be reliably calculated since the results are very sensitive to variation of the nuclear parameters.

The suppression of the matrix elements of W, d_z and S_z can be explained qualitatively. First, the angular momenta I of the dominant components in the Nilsson functions of the anomalously close states $|\Omega\rangle$ and $|\bar{\Omega}\rangle$ differ by two, and therefore are not mixed by the above operators.

Second, the density and potential have similar profiles in nonspherical nuclei as well. Then, if the spin–orbit interaction is neglected, we obtain, according to formula (10.8),

$$W(\boldsymbol{r}) = \mathrm{i}\,\lambda\,[\boldsymbol{\sigma} \cdot \boldsymbol{p},\, H_0],$$

where $H_0 = p^2/2m + U$ is the single-particle Hamiltonian. Therefore,

$$\langle \Omega | W | \bar{\Omega} \rangle = \mathrm{i}\lambda \langle \bar{\Omega} | [\boldsymbol{\sigma p}, H_0] | \Omega \rangle \propto E_\Omega - E_{\bar{\Omega}}, \tag{10.28}$$

and the small denominator in formula (10.27) cancels.

A similar suppression takes place for the matrix element of d_z in the Nilsson oscillator model. Indeed, if the potential is

$$U = \frac{1}{2} m_{\mathrm{p}} \left[\omega_\perp^2 (x^2 + y^2) + \omega_z^2 z^2 \right],$$

then

$$d_z = e\,z = \frac{e}{m_{\mathrm{p}} \omega_z^2} \frac{\partial U}{\partial z} = \frac{\mathrm{i}\,e}{m_{\mathrm{p}} \omega_z^2} [p_z, H_0], \tag{10.29}$$

and $\langle \Omega | d_z | \bar{\Omega} \rangle$ is also proportional to the small energy difference $E_\Omega - E_{\bar{\Omega}}$.

Thus, if relations (10.28) and (10.29) were exact, the contribution of the anomalously close level to S_z and M_{zz} would not be enhanced, and the contribution to d_z would even be suppressed. Actually, of course, this is not the case. The spin-orbit interaction cannot be neglected, density and potential profiles are not exactly similar, and the oscillator potential is an approximation only.

It is no wonder therefore that numerical calculations in the Nilsson model lead to somewhat different conclusions. The EDM is not enhanced in ^{237}Np and is possibly enhanced by 5 – 10 times in ^{161}Dy (the calculation is unstable). There is no enhancement of the Schiff moment in either nuclei. The magnetic quadrupole moment is enhanced in ^{161}Dy and ^{237}Np by approximately one order. These results are collected in Table 10.1 which also includes the predictions of [289, 327] for those spherical nuclei that are of experimental interest at present. It should be pointed out that many-body corrections to the results for ^{129}Xe, ^{133}Cs, ^{199}Hg, and ^{203,205}Tl are far from being negligible; for the last two nuclei they can reach 100%, while for the deformed

Table 10.1. Schiff moments and MQMs for several nuclei

	$S[e\ \mathrm{fm}^3] \times 10^7$	$M[(e/2m_{\mathrm{p}})\ \mathrm{fm}] \times 10^7$
^{129}Xe	$2.2\,\xi_{np}$	
^{133}Cs	$3.8\,\xi_p$	$3.6\,\xi_p$
^{199}Hg	$-1.8\,\xi_{np}$	
^{203,205}Tl	$1.5\xi_{pp} - 1.8\xi_{pn}$	
^{161}Dy	$\approx 1\,\xi_n$	$\approx 50\,\xi_n$
^{237}Np	$\approx 5\,\xi_p$	$\approx 40\,\xi_p$

nuclei ^{161}Dy and ^{237}Np, the predictions are no more than order-of-magnitude estimates.

In deformed nuclei there is one more enhancement mechanism, specific to the magnetic quadrupole [269]. It is well known that, due to collective effects, electric quadrupole moments of these nuclei are regularly larger than the corresponding single-particle predictions. The enhancement factor can be estimated as

$$\approx \beta_2 Z,$$

where the order of magnitude for the quadrupole deformation is (see [326], Vol. 2, Chaps. 4 and 6)

$$\beta_2 \approx A^{-1/3}.$$

It is only natural to expect that the same enhancement should take place for the MQM as well. This simple estimate is in good agreement with calculations for ^{163}Dy and ^{161}Dy [269] which give enhancement factors of about 5 to 6.

A specific enhancement is possible for the Schiff moment as well, but in nonspherical nuclei with an octupole deformation [330]. There are reasons to believe that some nuclei in the regions Ba – Sm and Ra – Th indeed possess such a pear-like deformation [331]. Of course, if CP is conserved, all T-odd moments of such a nucleus average to zero in a stationary state, with a fixed rotational angular momentum, in exactly the same way as molecular T-odd moments (Sect. 1.3). However, in the body-fixed system the dipole moment and the Schiff moment, generally speaking, should not vanish. Obviously, neither of these vector characteristics can be induced by an octupole deformation alone: Y_{10} cannot correlate with Y_{30}. But it can correlate with $Y_{30} \times Y_{20}$. In other words, both dipole and Schiff moments arise under combined action of the quadrupole and octupole deformations. Still, if one assumes that protons and neutrons are distributed alike, in other words, that the mass and charge densities are proportional, the nuclear dipole moment vanishes even in the body-fixed system. The value of the Schiff moment in the frozen system can be estimated as

$$S_{\text{int}} \sim \langle \boldsymbol{r} r^2 \rangle \sim 4\pi e R^3 \beta_2 \beta_3 Z. \tag{10.30}$$

At the typical values of the quadrupole and octupole deformations, $\beta_2 \approx 0.2$, $\beta_3 \approx 0.1$, the collective factor $\beta_2 \beta_3 Z$ is only about unity.

The possible enhancement lies elsewhere. The octupole deformation of the core admixes opposite parity states to the wave function of the odd, outer nucleon (as was the case with atomic orbitals for the electron in a heteronuclear molecule). A reasonable estimate for the magnitude of these admixtures is just β_3. The assumption is that the nuclear ground state and the first excited state are members of the parity doublet with the wave functions of the outer nucleon

$$\frac{1}{\sqrt{2}} \left(|\Omega\rangle \pm |-\Omega\rangle \right), \tag{10.31}$$

and with the same state of the core. In formula (10.31) Ω is the projection of the total angular momentum of the valence nucleon onto the core axis. The CP-odd interaction W mixes these opposite-parity states, creating in this way the nuclear Schiff moment S_{oct} in the laboratory system. The natural estimate for this mixing is now

$$\beta_3 \frac{\langle W \rangle}{\Delta E}, \tag{10.32}$$

where $\Delta E \sim 100$ keV is the energy interval between the doublet components. Now, using formulas (10.6), (10.8), (10.22), (10.30), and (10.32), we obtain the following ratio of thus induced Schiff moment to the single-particle moment S_1 of a spherical nucleus:

$$\frac{S_{\text{oct}}}{S_1} \sim \beta_2 \beta_3^2 Z \, \frac{U_0}{\Delta E}. \tag{10.33}$$

This ratio may exceed two orders of magnitude. But for such an enhancement to arise, two nontrivial conditions are crucial: the existence of the octupole deformation in the ground state, and the simple structure (10.31) of the parity doublet. It is not exactly clear whether these conditions hold in reality.

10.4 Theoretical Predictions and Implications

We will begin this discussion, as in the case of the neutron EDM, with the θ term contribution to the CP-odd nuclear forces. It has been demonstrated in Chap. 4 that this term (5.3) in the QCD Lagrangian generates the CP-odd πNN interaction with the coupling constant $\bar{g}_{\pi NN} \approx -0.027\theta$. There are strong reasons to believe that the CP-odd NN interaction is dominated in this case by the π^0 exchange. This mechanism, considered first in [329], is characterized by the large value, 13.6, of the strong πNN constant and by the small π meson mass. The derivative occurring at one of the vertices (in this case at the strong one) arises inevitably in the case of a P-odd interaction and does not lead to a relative suppression of the corresponding contribution. Finally, charged particle exchange is suppressed as compared to neutral exchange in the nuclear shell model.

To simplify the discussion, we will restrict ourselves to the limit of zero momentum transfer in the NN interaction. In this way the π^0 exchange induces an effective operator

$$\frac{G}{\sqrt{2}} \xi \, (\bar{N} \mathrm{i} \gamma_5 N)(\bar{N}' N') \tag{10.34}$$

with a dimensionless constant

$$\xi = \frac{g_{\pi NN} \, \bar{g}_{\pi NN} \sqrt{2}}{G \, m_\pi^2}. \tag{10.35}$$

In the nonrelativistic reduction, expression (10.34) in the case of a heavy nucleus becomes interaction (10.3) with the constant ξ averaged over the core nucleons.

Combining this estimate with expression from Table 10.1 for the ^{199}Hg Schiff moment and with the experimental result (8.92), we get an upper limit for the CP-odd QCD parameter comparable to that from the neutron EDM (5.11):

$$|\theta| < 7 \times 10^{-10}. \tag{10.36}$$

The same result can be formulated as an upper limit on the effective CP-odd neutral pion constant:

$$\bar{g}^0_{\pi NN} < 2 \times 10^{-11}. \tag{10.37}$$

Let us now consider the standard model predictions for CP-odd nuclear forces. It was pointed out in [278] that this interaction is most efficiently mediated by the neutral K meson exchange. However, values given in [278], for the concrete mechanism considered there, were essentially overestimated [195, 332]. A new (and, hopefully, truly dominant) type of the kaon exchange was suggested in [333].

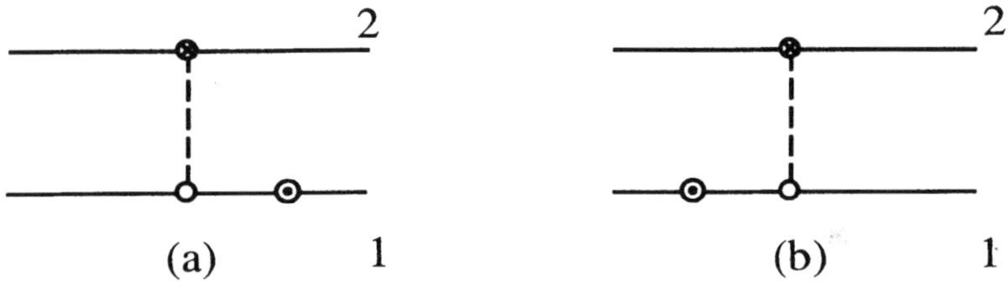

Fig. 10.1a,b. Meson exchange diagram

This exchange reduces essentially to the diagrams of Fig. 10.1. The CP-odd scalar K^0NN vertex $\otimes$ on the upper line is generated again by the "penguin" diagram. The lower line describes the pole contribution to the weak CP-even P wave amplitude. The strong meson–baryon vertex $\circ$ is well known from ordinary SU(3) symmetry. The weak baryon–baryon transition $\odot$ is found from the baryon pole fit to the usual P wave nonleptonic hyperon decay amplitude. Under the usual octet dominance assumption, the weak baryon–baryon transition can be written as follows (see, e.g., [150]):

$$a_1(S\bar{B}B) + a_2(SB\bar{B}).$$

Here $S = \lambda_6$ and B is the baryon octet SU(3) matrix. A reasonable fit to the four independent P-wave amplitudes, known experimentally (see Table 10.2 where we follow the sign convention for the amplitudes adopted in [334]), is obtained with

Table 10.2. Fit to P wave amplitudes for nonleptonic hyperon decays

	Λ^0_-	Σ^+_0	Σ^-_-	Ξ^-_-
our fit	–11.3	–10.5	1.3	–9.0
experimental value	–9.98(24)	–12.04(58)	0.65(7)	–7.49(28)

$$a_1 = 0.75\, m_\mathrm{p}, \quad a_2 = -0.097\, m_\mathrm{p}.$$

In this way, we arrive at the following set of CP-odd nucleon–nucleon constants (assuming for the combination of the Kobayashi–Maskawa angles the numerical value $\sin\delta s_1 s_2 s_3 = 2 \times 10^{-4}$):

$$\xi_{pp} = -0.9 \times 10^{-9}, \quad \xi_{pn} = -1.1 \times 10^{-9},$$
$$\xi_{np} = 1.2 \times 10^{-9}, \quad \xi_{nn} = 1.4 \times 10^{-9}. \tag{10.38}$$

These values of the CP-odd constants [335] agree with those derived in [333] for a valence neutron, i.e., for the constants ξ_{np}, ξ_{nn}. There is however a qualitative discrepancy concerning constants ξ_{pp}, ξ_{pn}; the assertion of [333] is that they are smaller by two orders of magnitude.

The next model is that with spontaneous CP violation in the Higgs sector. Its old version, with light Higgs bosons, has been ruled out by the experimental upper limits on the neutron EDM. We will consider therefore a more "natural" version, with heavy Higgs bosons. As it was the case with the neutron dipole moment, the most meaningful limits follow here from considerations of the quark chromoelectric dipole moments. The CP-odd $\pi^0 NN$ vertex generated by the quark CEDM transforms by the PCAC technique:

$$\langle\, \pi^0 N \,|\, g\, \bar{q}\gamma_5 \sigma_{\mu\nu} t^a q\, G^a_{\mu\nu} |\, N\,\rangle$$
$$= \pm\frac{\mathrm{i}\sqrt{2}}{f_\pi}\, \langle\, N \,|\, g\, \bar{q}\sigma_{\mu\nu} t^a q\, G^a_{\mu\nu}\,|\, N\,\rangle, \tag{10.39}$$

where the plus and minus refer to the u and d quark CEDM, respectively. The QCD sum rule estimate for the last expectation value is 7 GeV^2 [195]. In this way we obtain, in the model discussed, the following prediction for the CP-odd πNN constant:

$$\bar{g}^0_{\pi NN} < 2.5 \times 10^{-10}. \tag{10.40}$$

This estimate, obtained in the "natural" version of the model, under the assumption that the Higgs mass is about the same as that of t quark, is 12 times larger than the experimental upper limit (10.37) [191]. The same situation occurs in the supersymmetric SO(10) model [194].

Thus, the atomic experiment constrains the parameters of both model even more strongly than the upper limit on the neutron EDM.

To conclude this section, we once more adopt a phenomenological approach. We will now present upper limits, following from atomic experiments, on the effective operators discussed previously in connection with the neutron EDM measurements. Most of these limits are obtained under the assumption that CP-odd nuclear forces are mediated by π^0 exchange. In other words, the problem is reduced here to the calculation of the CP-odd neutral pion constant $\bar{g}^0_{\pi NN}$ constrained experimentally by (10.37). The exception is the operators characterized by the constants k^c_s at $q_1 \neq q_2$, k^c_t, and k^g; the CP-odd $\pi^0 NN$ vertex is rather insensitive to these. Therefore, the limits for these operators following from the atomic experiment have been derived from the bound (8.93) on the neutron EDM extracted from the same mercury result. All the results are summarized, together with those derived in Chap. 5 from the neutron experiments, in Table 10.3.

These limits on the effective CP-odd interaction operators are quite impressive. All the constants are several orders of magnitude less than the usual Fermi weak interaction constant G.

Table 10.3. Limits on various operators as deduced from EDM experimental results

	$d(\text{n})/e < 10^{-25}\,\text{cm}$	$d(^{199}\text{Hg})/e < 9 \times 10^{-28}\,\text{cm}$
$\frac{G}{\sqrt{2}} k_s (\bar{q_1} \mathrm{i}\gamma_5 q_1)(\bar{q_2} q_2)$	$\lvert k_s \rvert < 3 \times 10^{-5}$	$\lvert k_s \rvert < 2 \times 10^{-6}$
$\frac{G}{\sqrt{2}} k^c_s (\bar{q_1} \mathrm{i}\gamma_5 t^a q_1)(\bar{q_2} t^a q_2)$		
$q_1 = q_2$	$\lvert k^c_s \rvert < 10^{-4}$	$\lvert k^c_s \rvert < 7 \times 10^{-6}$
$q_1 \neq q_2$	$\lvert k^c_s \rvert < 10^{-4}$	$\lvert k^c_s \rvert < 6 \times 10^{-4}$
$\frac{G}{\sqrt{2}} k_t \frac{1}{2} \epsilon_{\mu\nu\alpha\beta} (\bar{u}\sigma_{\mu\nu} u)(\bar{d}\sigma_{\alpha\beta} d)$	$\lvert k_t \rvert < 10^{-5}$	$\lvert k_t \rvert < 7 \times 10^{-5}$
$\frac{G}{\sqrt{2}} k^c_t \frac{1}{2} \epsilon_{\mu\nu\alpha\beta} (\bar{u}\sigma_{\mu\nu} t^a u)(\bar{d}\sigma_{\alpha\beta} t^a d)$	$\lvert k^c_t \rvert < 10^{-5}$	$\lvert k^c_t \rvert < 5 \times 10^{-5}$
$[d^c(q)/2]\bar{q}\gamma_5 \sigma_{\mu\nu} g G^a_{\mu\nu} t^a q$	$\lvert d^c(q) \rvert < 10^{-25}\,\text{cm}$	$\lvert d^c(q) \rvert < 2.4 \times 10^{-26}\,\text{cm}$
$[d^c(s)/2]\bar{s}\gamma_5 \sigma_{\mu\nu} g G^a_{\mu\nu} t^a s$	$\lvert d^c(s) \rvert < 1.4 \times 10^{-24}\,\text{cm}$	$\lvert d^c(s) \rvert < 1.7 \times 10^{-25}\,\text{cm}$
$\frac{G}{\sqrt{2}} k^g \frac{1}{6} \epsilon_{\mu\nu\alpha\beta} f^{abc} G^a_{\mu\nu} G^b_{\alpha\rho} G^c_{\beta\rho}$	$\lvert k^g \rvert < 4 \times 10^{-5}$	$\lvert k^g \rvert < 3 \times 10^{-4}$
$\theta \frac{\alpha_s}{8\pi} \frac{1}{2} \epsilon_{\mu\nu\alpha\beta} G^a_{\mu\nu} G^a_{\alpha\beta}$	$\lvert \theta \rvert < 3 \times 10^{-10}$	$\lvert \theta \rvert < 7 \times 10^{-10}$

11. What Do We Really Know About T-Odd, but P-Even Interactions?

By mere exertion of a will ...

M. Tsvetayeva

Direct experimental information about T-odd, but P-even (TOPE) interactions is rather scant. The best limits on the relative magnitude of the corresponding admixtures to nuclear forces lie around 10^{-3} [336–339]. In the following, we will relate all interactions to the Fermi weak interaction constant G. Since the nuclear scale of weak interactions is $Gm_\pi^2 \approx 2 \times 10^{-7}$, those limits can be formulated as $10^4 G$. The most advanced experimental proposals aim at improving these limits by three orders of magnitude.

Experimental information on TOPE electron–nucleon interactions is practically absent. In [340] an atomic experiment was suggested which can hopefully reach an accuracy $\approx 3 \times 10^4 G$ (see also [341]). Higher accuracy is aimed at in the experimental proposal [342].

As to a TOPE electron–electron interaction, its possible manifestations in positronium were discussed in [343].

Much better upper limits on the TOPE interactions can be obtained as follows. Radiative corrections, due to the P-odd part of the electroweak interaction, transform the T-odd, but P-even fermion–fermion interaction into a T-odd and P-odd interaction. The experimental information about T-odd, P-odd effects is sufficiently rich to obtain in this way new limits, much better than direct ones, on the parameters of T-odd, P-even electron–electron, electron–nucleon, and nucleon–nucleon interactions.

11.1 Long-Range Effects

Let us first point out that the predictions of all modern renormalizable theories of CP violation (and not only the standard model!) cannot exceed $(10^{-3} - 10^{-4})\, G$. The reason is obvious. Parity violation is an intrinsic property of all these models, and therefore T-odd, P-even effects can be expected to be roughly of the same order of magnitude as T-odd, P-odd effects.

An even stronger result was obtained recently in [344]. In any renormalizable theory, TOPE flavor conserving quark–quark interactions are absent to second order in the boson–fermion and boson–boson coupling. This conclusion holds to all orders in both the chromodynamic and the electromagnetic interactions, provided the θ term is neglected.

However, in no way does it mean by itself that the experimental efforts in this field do not make sense. They do, but it should be understood clearly that this is the search for essentially new physics, well beyond the modern theories. That is why we resort again to the phenomenological approach, using the TOPE quark–quark operators (2.16).

An idea of the kind of limits that can be obtained by combining the interaction discussed above with the P-odd interaction is given by the following argument, close in spirit to the corresponding estimates from [345, 346]. Let us consider the contribution to the neutron EDM from the combined action of the usual P-odd, T-even weak interaction and the discussed T-odd and P-even interaction, the strength of the latter being q times smaller than that of the former. The contribution obviously amounts to

$$d_{\mathrm{n}}/e \approx \frac{1}{m_{\mathrm{p}}}(Gm_{\pi}^2)^2 q. \tag{11.1}$$

From a comparison with the experimental result (4.43) we obtain the limit $q < 10^2$, which is about two orders of magnitude better than the direct limits mentioned above. This estimate is obviously of a very crude nature. In particular, the dipole moment arises here at least in a one-loop approximation which leads to a small geometrical factor. So, it is better perhaps to accept for this limit a more cautious estimate

$$q < 10^2 - 10^3, \tag{11.2}$$

which can alternatively be formulated as the upper limit

$$\alpha_{\mathrm{T}} < 10^{-5} - 10^{-4} \tag{11.3}$$

for the relative magnitude α_{T} of the TOPE admixture to nuclear forces.

Analogous estimates for the electron–nucleon interaction were made in [347, 348] (as cited in [349]) and [350]. Let us mention also recent elaborate investigations [351–354] of the long-distance interplay between TOPE and the usual P-odd interactions in the hadronic sector. They are of a certain interest for theoretical nuclear physics, but none resulted in a serious improvement over the simple estimate (11.3).

Better limits on TOPE effects are obtained using a simple and elegant argument presented in [355]. Through dimensional reasoning, the TOPE four-fermion effective interaction of dimension seven can be written as

$$\frac{C_7}{\Lambda^3}\bar{\psi}_1 \mathrm{i}\gamma_5 \sigma_{\mu\nu}(p_1' - p_1)_\nu \psi_1 \bar{\psi}_2 \gamma_\mu \gamma_5 \psi_2. \tag{11.4}$$

It is only natural to assume that the momentum scale Λ in this operator exceeds that of the electroweak theory, i.e.,

$$\Lambda > 100 \text{ GeV}.$$

It is natural to assume that the dimensionless number C_7 is about unity.

Then the dimensional estimate for the magnitude of TOPE effects on the hadronic scale of momenta $p \approx 1$ GeV is

$$\left(\frac{p}{\Lambda}\right)^3 < 10^{-6}. \quad (11.5)$$

This line of reasoning applies not only to four-fermion operators, but, for instance, to a quark–gluon–photon operator of the form [355]

$$\frac{C_7'}{\Lambda^3}\bar{q}\sigma_{\mu\nu}t^a q\, G^a_{\mu\rho}F_{\nu\rho}. \quad (11.6)$$

Much stronger constraints follow in this way for TOPE photon–fermion scattering amplitudes, which belong to higher dimensions (starting from 10) [356].

11.2 TOPE Fermion–Fermion Interactions. One-Loop Approach

Significant progress in tackling this problem came with the observation that the electroweak corrections to the operators (2.16) are controlled mainly not by the long-distance effects, but by short-distance ones. Therefore, they are of the order α/π (up to some chiral suppression factor which is quite essential), but not of the order Gm_π^2 [349].

We will concentrate (here and below) on the corrections primarily due to the Z boson exchange. These can be calculated self-consistently in the sense that the result is independent of the choice of gauge for the Z boson propagator. The result is also independent of the form of the initial T-odd operators (2.16), which can be rewritten as

$$\bar{\psi}_1 \mathrm{i}\gamma_5(p_1' + p_1)_\mu \psi_1 \bar{\psi}_2 \gamma_\mu \gamma_5 \psi_2,$$

$$\bar{\psi}_1 \gamma_\mu \gamma_5 \psi_1 \bar{\psi}_2 \mathrm{i}\gamma_5 (p_2' + p_2)_\mu \psi_2. \quad (11.7)$$

The representation (11.7) is more convenient for most calculations.

A consistent, gauge-independent calculation of the W boson exchange contribution to the induced T- and P-odd amplitudes is much more model-dependent and will not be discussed here in detail. It can be expected however to be even larger than that of the Z exchange, due to the small numerical values of the neutral weak charges. These small values are responsible in particular for the well known relative suppression of the neutral-current cross-sections as compared to those of the charged-current. So, the Z-exchange contribution serves as an estimate from below for the effects discussed. As for Higgs boson exchange, in the standard model it conserves parity and is therefore of no interest to us.

We therefore take the TOPE operator as

$$\frac{G}{\sqrt{2}}\frac{q_1}{2m_p}\,\bar{\psi}_1 i\gamma_5(p_1'+p_1)_\mu\psi_1\bar{\psi}_2\gamma_\mu\gamma_5\psi_2,$$

As was mentioned above, we measure the interaction discussed in the units of the Fermi weak interaction constant G; m_p is the proton mass, its choice as the necessary dimensional parameter being also a matter of convention; q_1 is dimensionless.

The result of transforming this effective operator into a T- and P-odd one is

$$\frac{G}{\sqrt{2}}\frac{\alpha}{3\pi}\log\frac{\Lambda^2}{M^2}\frac{q_1}{m_p}\,\{2m_2v_1[3a_2\bar{\psi}_1\psi_1\,\bar{\psi}_2\,i\gamma_5\,\psi_2-(a_1+a_2)\bar{\psi}_1\psi_1\bar{\psi}_2 i\gamma_5\psi_2]$$

$$-a_1v_2[m_2\bar{\psi}_1 i\gamma_5\sigma_{\mu\nu}\psi_1\,\bar{\psi}_2\sigma_{\mu\nu}\psi_2-\bar{\psi}_1 i\gamma_5\sigma_{\mu\nu}(p_1'-p_1)_\nu\psi_1\,\bar{\psi}_2\gamma_\mu\psi_2]\}. \quad (11.8)$$

Here M is the Z boson mass. The dependence of the result on the cut-off parameter Λ is due to the nonrenormalizability of the TOPE interaction. But trying to be as conservative as possible in our numerical estimates, we will assume the log to be on the order of unity. $m_{1,2}$ are the masses of the first and second fermions, respectively. $v_{1,2}$ and $a_{1,2}$ are their weak neutral vector and axial charges. In particular, for the electron, u , and d quarks they are:

$$\begin{aligned} v_e &= -\tfrac{1}{2}(1-4\sin^2\theta)\approx -0.04, & a_e &= -\frac{1}{2},\\ v_u &= \tfrac{1}{2}(1-\tfrac{8}{3}\sin^2\theta)\approx \tfrac{1}{6}, & a_u &= \frac{1}{2},\\ v_d &= -\tfrac{1}{2}(1-\tfrac{4}{3}\sin^2\theta)\approx -\tfrac{1}{3}, & a_d &= -\frac{1}{2}. \end{aligned} \quad (11.9)$$

Here θ is the electroweak mixing angle,

$$\sin^2\theta\approx 0.23.$$

Let us perform the concrete estimates for the electron–nucleon interaction. In this case the induced electron–quark operator is

$$\frac{G}{\sqrt{2}}\frac{\alpha}{3\pi}\log\frac{\Lambda^2}{M^2}\,v$$

$$\left\{q_{eq}\left[\frac{m}{m_p}\,\bar{e}\,i\gamma_5\sigma_{\mu\nu}e\,\bar{q}\,\sigma_{\mu\nu}q-\frac{1}{2m_p}\,\bar{e}\,i\gamma_5\sigma_{\mu\nu}(p_1'-p_1)_\nu e\,\bar{q}\,\gamma_\mu q\right]\right.$$

$$\left.+\,q_{qe}\frac{m_e}{m_p}\,[(1-2a)\,\bar{e}\,i\gamma_5 e\,\bar{q}\,q-3\,\bar{e}\,e\,\bar{q}\,i\gamma_5 q]\right\}. \quad (11.10)$$

Here m_q and m_e are the quark and electron masses, v and a are the quark vector and axial charges, q_{eq} and q_{qe} are the dimensionless constants in the T-odd, P-even operators with the explicit momenta belonging to electrons and quarks, respectively. We have neglected here the contribution proportional to the electron vector charge v_e which is numerically small [see (11.9)]. Operator (11.10) should be summed over u and d quarks, and its expectation value should be taken first over a nucleon and then over a nucleus.

In the static approximation for nucleons, the only term in (11.10) that depends on both electron and nucleon spin is $\gamma_5\sigma_{\mu\nu} \times \sigma_{\mu\nu}$. The dimensional estimate for the nucleon expectation value of the operator $\bar{q}\sigma_{\mu\nu}q$ is $\bar{N}\sigma_{\mu\nu}N$. Then the dimensionless effective constant of the T- and P-odd tensor electron–neutron interaction is

$$k_2 = \frac{\alpha}{3\pi} \log \frac{\Lambda^2}{M^2} \frac{m_{\mathrm{d}}}{2m_{\mathrm{p}}} v_d q_{eq} \approx 10^{-6} q_{eq}. \tag{11.11}$$

The upper limit on the constant k_2 (8.43) leads to the following result:

$$q_{eq} < 4 \times 10^{-2}. \tag{11.12}$$

For the d quark mass we assume here the value $m_{\mathrm{d}} = 7$ MeV.

Upper limits on the order of 0.1 are extracted in this way from atomic experiments for the constants q_{qe} referring to the electron–quark interaction with the derivatives in the quark vertex.

The analogous estimates for the quark–quark constants are more tedious, due mainly to the problem of evaluating the contributions of the arising T- and P-odd four quark operators to the neutron EDM. In this way one gets for different quark–quark constants upper limits at the level

$$q_{qq} < 10. \tag{11.13}$$

We do not go into details here since in the next section much better upper limits will be obtained for all these constants.

11.3 TOPE Fermion–Fermion Interactions. Two-Loop Approach

The idea of the next improvement of the upper limits on the constants discussed [357, 358] can be conveniently explained for the case of hadrons. The previous advance from (11.2) to (11.13) was obtained by going from the long-distance effects of the usual weak interaction to short-distance effects. This allowed us to get rid of one small factor Gm_π^2 in formula (11.1), trading it for α/π with some extra chiral suppression. But can we get rid of the second factor Gm_π^2 in that formula? The answer is: Yes, we can. Up to now we computed the one-loop radiative correction which transformed the interaction discussed into a T- and P-odd effective operator. Then we estimated in fact the long-distance contribution of this operator to the neutron dipole moment. Now we are going to make the next step: To calculate an exclusively short-distance two-loop contribution of the TOPE interaction times the weak interaction, directly to the quark EDM. And the latter at least does not significantly exceed the neutron dipole moment. In this way we gain even more than at the first step since now there is no more chiral suppression factor m_q/m_p.

To regularize, at least partly, the related Feynman integrals, it is convenient to introduce explicitly an axial boson of mass μ mediating the TOPE interaction. It was pointed out long ago [359] (see also [360–362]) that the amplitudes (2.16) could arise through the exchange by a neutral pseudovector boson if its vertices contain a mixture of the "normal" axial operator $\gamma_\mu\gamma_5$ and the "anomalous" one $i\gamma_5\sigma_{\mu\nu}(p'-p)_\nu$ of opposite CP-parity. We consider however the introduction of this boson only as a convenient way to soften the ultraviolet divergence of the integral from a quadratic to a logarithmic one, without any serious discussion of this particle itself. Thus the TOPE fermion–fermion amplitude can be presented as

$$\frac{4\pi\beta}{\mu^2-k^2}\,\frac{1}{2m_p}\,\bar{\psi}_1 \mathrm{i}\gamma_5\sigma_{\mu\nu}(p_1'-p_1)_\nu\psi_1\,\bar{\psi}_2\gamma_\mu\gamma_5\psi_2. \tag{11.14}$$

Here β is the dimensionless coupling constant analogous to $\alpha = 1/137$ in QED.

Consider now the contribution of the two-loop diagram (a) in Fig. 11.1 to the fermion EDM. Here the dashed line represents the propagator of the axial boson X. The Z boson exchange (the wavy line) introduces the parity nonconservation necessary to induce the EDM.

It can be seen easily that the contribution proportional to the mass of fermion 1 is proportional to G. Therefore, this fermion will be taken as massless. Since the dipole moment interaction (2.5) changes the fermion chirality, the lower vertex of the X line should be $\mathrm{i}\gamma_5\sigma_{\mu\lambda}k_\lambda$ and the upper one, correspondingly, $\gamma_\mu\gamma_5$. Then according to the Furry theorem for the fermion loop, the upper vertex of the Z line is γ_ν; therefore the lower one is $\gamma_\nu\gamma_5$. To simplify both the calculations and the result, we will neglect the mass of the fermion propagating in the upper loop as well (though for the heavy t quark there are no special reasons to do so). Then the expression for the fermion loop is [169]

$$\frac{\mathrm{i}}{8\pi^2}F_{\alpha\beta}\left[\epsilon_{\alpha\beta\mu\nu}+\frac{1}{k^2}\epsilon_{\alpha\beta\kappa\lambda}k_\kappa(\delta_{\mu\lambda}k_\nu+\delta_{\nu\lambda}k_\mu)\right]; \tag{11.15}$$

here $F_{\alpha\beta}$ is the strength of the external field.

When considering the lower (Compton) block of diagram (a) in Fig. 11.1, one should also include the contact term

$$\frac{e}{2\sin\theta\cos\theta}\,2\,a\,\mathrm{i}\,\sigma_{\mu\nu}, \tag{11.16}$$

originating from the vertex

$$\mathrm{i}\gamma_5\sigma_{\mu\lambda}(p'-p)_\lambda$$

via the substitution

$$p_\mu\psi \to \left[p_\mu - \frac{e}{2\sin\theta\cos\theta}\,(v+a\gamma_5)\,Z_\mu\right]\,\psi,$$

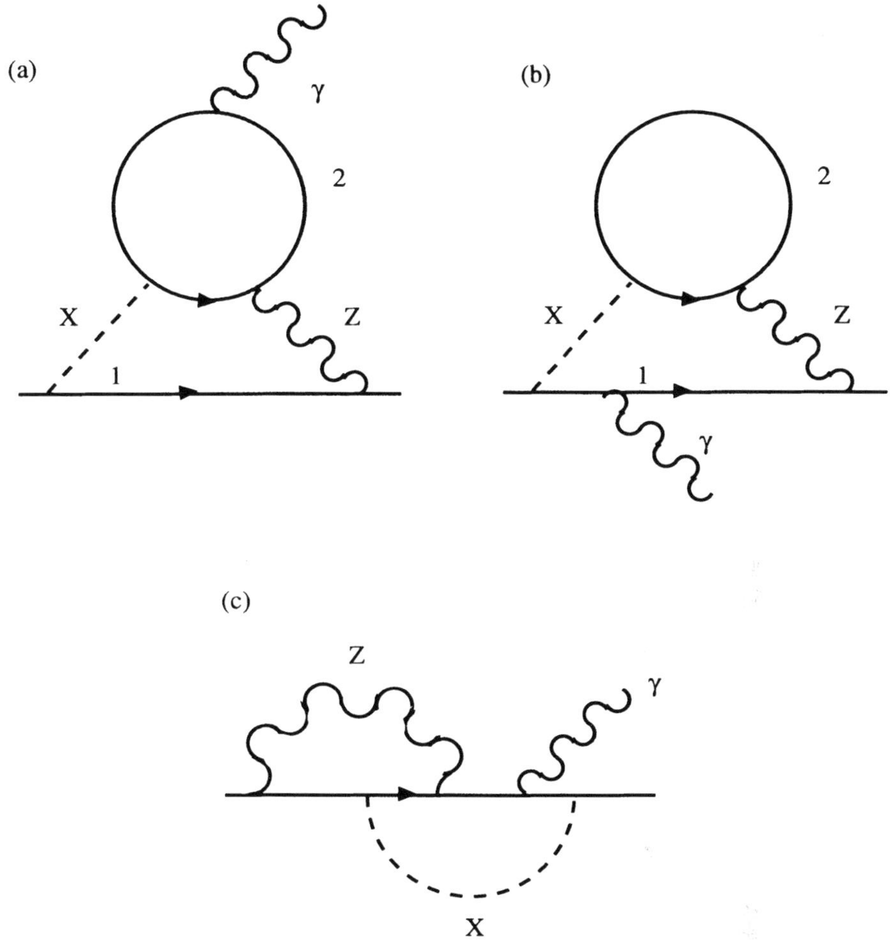

Fig. 11.1a–c. Two-loop diagrams

which makes this vertex gauge invariant with respect to the Z field. In particular, the inclusion of the contact vertex (11.16) into the XZ Compton scattering amplitude makes the result of the calculation independent of the term $k_\nu k_\sigma / M^2$ in the propagator of the Z boson. Simple calculations give the following result for this contribution to the EDM d of fermion 1:

$$\frac{d}{e} = -\frac{\alpha \beta Q_2 a_1 v_2 N}{3\pi^2 m_{\mathrm{p}}} \log \frac{\Lambda^2}{M_>^2} . \qquad (11.17)$$

This formula refers to the general case in which fermion 2 propagating in the loop differs from the fermion 1 propagating in the lower line. If fermion 2 is a quark, $N = 3$ (number of colors), and if it is an electron, $N = 1$; a_1 is the axial weak charge of the first fermion; v_2 is the vector weak charge of the second fermion, and Q_2 is its electric charge in the units of e. The logarithmic dependence on the cut-off parameter Λ is the result of the nonrenormalizable coupling of the X boson to the vertex with derivatives $i\gamma_5\sigma_{\mu\rho}k_\rho$. Although the result (11.17) is presented with logarithmic accuracy, in all of our numerical estimates we will conservatively assume $\log \Lambda^2/M_>^2$ to be of the order of unity ($M_>$ is the larger of the masses μ and M). Let us emphasize again the gauge invariance of this result with respect to the Z field. Its gauge invariance with respect to the electromagnetic field is self-evident from expression (11.15).

The contribution of diagram (b) of Fig. 11.1,

$$\frac{d_b}{e} = \frac{\alpha\,\beta\,Q_1 a_2 v_1 N}{36\pi^2 m_{\rm p}} \log^2 \frac{\Lambda^2}{M_>^2}, \tag{11.18}$$

is much smaller numerically and can be neglected.

In the case of identical fermions there is also the contribution to the EDM of diagram (c), but we will neglect it in our estimates with the expectation that the result will not be grossly affected. One might expect here that in the local limit, $\mu \to \infty$, the effect for identical fermions 1 and 2 should vanish, which would correspond to exact cancelation of diagrams (a) and (c) of Fig. 11.1. However, even for $\mu \gg M$ we can, to logarithmic accuracy, restrict the integration over k to $k \gg \mu$, where the TOPE interaction of identical fermions is in no way a local one and therefore no cancellation takes place.

A consistent, gauge-independent calculation of the W boson exchange contribution to the induced EDM is again much more model dependent and we will not discuss it. The same arguments as in the previous section lead us to expect that the Z boson contribution serves as a conservative estimate for the induced EDM.

We will start the application of the general result (11.17) with the case of the electron–electron TOPE interaction. Substituting into formula (11.17) the numerical values (11.9) for a_e and v_e, as well as $Q_e = -1$, we get

$$d_e \approx \beta_{ee} \times 3 \times 10^{-19}\ e\,\text{cm}. \tag{11.19}$$

The upper limit (6.5) on the electron EDM leads to the following result for the constant β_{ee} of the electron–electron TOPE interaction:

$$\beta_{ee} < 10^{-8}. \tag{11.20}$$

In the same way we can get very strict upper limits on the electron–nucleon and nucleon–nucleon TOPE interactions. The axial charge of a fermion is always (up to a sign) 1/2 and for any quark, independent of its type, the product Qv is numerically close to 1/9. Then, using the experimental upper limit (4.43) on the neutron EDM and assuming for dimensional reasons that the neutron dipole moment induced by the quark EDM is of about the same magnitude as the latter, we get for β_{qe}, the TOPE quark–electron interaction constant with derivatives in the quark vertex, the limit

$$\beta_{qe} < 3 \times 10^{-7}. \tag{11.21}$$

For another electron–quark constant β_{eq} (with the derivative in the electron vertex) the constraint (6.5) on the electron EDM gives the upper limit

$$\beta_{eq} < 10^{-9}. \tag{11.22}$$

For all quark–quark constants β_{qq} the limit (4.43) on the neutron EDM gives

$$\beta_{qq} < 10^{-8} - 10^{-7}. \tag{11.23}$$

The latter also refers to the "colored" TOPE interaction, with the $SU(3)$ generators t^a in each vertex. In this case the external field on diagram in Fig. 11.1 should not be electromagnetic, but rather a gluon one. Again, for dimensional reasons, the neutron EDM induced in this way should be of the same order of magnitude as the chromoelectric dipole moment described by this diagram.

The constants β introduced here, are related as follows to the constants q used above:

$$\frac{4\pi\beta}{\mu^2} = \frac{G}{\sqrt{2}}q, \tag{11.24}$$

or

$$q = 4\pi\beta\sqrt{2}\left(\frac{m_{\mathrm{p}}}{\mu}\right)^2 \times 10^5 = 1.8\times 10^6\left(\frac{m_{\mathrm{p}}}{\mu}\right)^2\beta. \tag{11.25}$$

Thus, the upper limits corresponding to (11.21), (11.22) and (11.23) are

$$q_{qe} < 0.5\left(\frac{m_{\mathrm{p}}}{\mu}\right)^2, \quad q_{eq} < 2\times 10^{-3}\left(\frac{m_{\mathrm{p}}}{\mu}\right)^2, \quad q_{qq} < 0.1\left(\frac{m_{\mathrm{p}}}{\mu}\right)^2, \tag{11.26}$$

respectively.

In the previous section we derived the upper limits on the constants q, and found values close to $q < 10$, under the assumption $\mu \geq M \approx 100 m_{\mathrm{p}}$. Under the same assumption, the limits we obtain here are much better:

$$q_{qe} < 10^{-4}, \quad q_{eq} < 10^{-7}, \quad q_{qq} < 10^{-5}. \tag{11.27}$$

Let us come back to the explanation of this improvement. In the transition from the effective four-fermion T- and P-odd operators obtained in the previous section to the neutron EDM, we used the usual hadronic scale of 1 GeV. But here the transition takes place on a much higher scale of 100 GeV.

11.4 Conclusions on TOPE eN and NN Interactions

Now, having obtained the above limits on the TOPE electron–quark and quark–quark interactions, what can we say about the corresponding electron–nucleon and nucleon–nucleon interactions?

The answer for the electron–nucleon interaction is quite straightforward. Simple dimensional arguments lead to the following estimates for the nucleon expectation values of the relevant quark operators (already mentioned in Sect. 8.3):

$$\langle N|\,\bar{q}\gamma_\mu\gamma_5 q\,|N\rangle \approx \bar{N}\gamma_\mu\gamma_5 N, \quad \langle N|\,\bar{q}\gamma_5\sigma_{\mu\nu}q\,|N\rangle \approx \bar{N}\gamma_5\sigma_{\mu\nu}N. \tag{11.28}$$

Therefore, the limits (11.27) for $q_{qe,eq}$ are readily translated into those for the constants of TOPE electron–nucleon interactions:

$$q_{Ne} < 10^{-4}, \qquad q_{eN} < 10^{-7}. \tag{11.29}$$

Let us now address the nucleon–nucleon interactions. Note first of all that in contrast to T- and P-odd nuclear forces, TOPE ones cannot be mediated by π^0-meson exchange [363]. Indeed, looking at the classification of the particle–antiparticle states in the annihilation channel presented in Sect. 2.3, we see that at $j = 0$ the state 2 just does not exist.

The absence of this exchange can be attributed also to the vanishing of a TOPE $\pi^0 NN$ vertex. The TOPE $\pi^\pm NN$ coupling, being hermitian, should be written as

$$\bar{p}\,\gamma_5 n\,\pi^+ - \bar{n}\,\gamma_5 p\,\pi^-. \tag{11.30}$$

This coupling does not lead to TOPE NN scattering amplitude in the one-boson exchange approximation since, after the interchange of this vertex and of the strong one, the corresponding diagrams cancel. TOPE one-boson exchange starts therefore with vector and pseudovector bosons. Being mediated by heavier particles, the effective NN interaction is further suppressed as compared to simple estimates.

On the other hand, it follows already from general formulas (2.16) that a TOPE nucleon–nucleon scattering amplitude contains an extra power of p/m_{p} as compared to the usual P-odd weak interaction. This means an extra suppression of by roughly an order of magnitude as compared with the above-mentioned naive estimate $Gm_\pi^2 q$.

Thus, even taking into account all the uncertainties in our estimates, one can state that the relative strength of the TOPE nuclear forces does not exceed $10^{-4}Gm_\pi^2$, or

$$\alpha_{\mathrm{T}} < 10^{-12}. \tag{11.31}$$

Various initial objections to this conclusion have died away, being withdrawn implicitly or explicitly. The only one still worth mentioning is the possibility that the contributions of various particles to the fermion loop in diagram (a) of Fig. 11.1 cancel out. However, this possibility, mentioned in [352], refers to any estimates (including those of [352]) made in the absence of a reliable theory. As for the analogy with the well known GIM mechanism mentioned in [352], it does not seem relevant here. The reasons for the GIM cancelation in the standard model are well known, but they also seem irrelevant to the issue of nonrenormalizable TOPE interactions. On the other hand, a too strong cancelation here seems especially unlikely, due to the large mass of the t quark. Moreover, a cancelation between contributions of different quarks at the level of 10^{-7}, which is required to change the upper limit 10^{-12} [see (11.31)] to 10^{-5} as discussed in [351–354], seems quite improbable.

There is however an essential inherent assumption, explicit or implicit, in all mentioned theoretical derivations of upper limits on TOPE interactions. It is assumed in all of them that the momentum scale related to the interactions is sufficiently high. In particular, the hypothetical axial boson mediator, should be very heavy. How can this assumption be substantiated, other than by invoking the usual paradigm according to which new physics should start at a new, higher momentum scale? First of all, numerous experimental searches for exotic light particles have, up to now, failed. On the other hand, it is at least nontrivial to accommodate a light (pseudo)vector boson, interacting with a nonconserved current, in a self-consistent theory.

Nevertheless, the final answer to the question of the existence of a (at least, relatively) long-range TOPE interaction can be given by experiment only. And this is the motivation for the atomic and positronium experiments discussed in [342, 358].

To conclude our discussion of the TOPE nuclear forces, it should be pointed out that their observable effects may considerably exceed the limit 10^{-12}; this is due to long-distance enhancement factors, such as small energy intervals between resonances mixed by this interaction.

11.5 T-Odd β Decay Constants

Some information about β decay constants can be obtained using an analogous approach [349, 357]. To relate them to the eN TOPE interaction, one should evidently switch on W exchange. As has been mentioned already, this procedure is more ambiguous than switching on Z exchange, as was done in our previous consideration. One can hope, however, that the estimates made below are valid, at least in their order of magnitude.

We start from the consideration of T-odd quark–lepton β decay interactions without derivatives. To obtain the radiative correction of the order of magnitude α/π, but not G, the quark mass should be neglected. Let us recall that all T- and P-odd fermion–fermion interaction operators without derivatives change the chirality of both fermions. Since the W exchange vertices contain left projectors, we can in this way, neglecting the quark masses, investigate only the chirality-changing quark–lepton operators:

$$(\bar{u}d)[\bar{e}(c_S + c'_S\gamma_5)\nu] + (\bar{u}\gamma_5 d)[\bar{e}(c_P\gamma_5 + c'_P)\nu] \\ - \frac{1}{2}(\bar{u}\sigma_{\mu\nu}d)[\bar{e}(c_T + c'_T\gamma_5)\sigma_{\mu\nu}\nu] + \text{h.c.} \tag{11.32}$$

So, in this approximation nothing can be said about axial and vector constants. On the other hand, the P-even part of the electroweak correction allows us to obtain information on the P- and T-odd β decay constants as well.

The effective T- and P-odd interaction of an electron with an u quark arising from (11.32) through W exchange is

$$\frac{G}{\sqrt{2}}\frac{\alpha}{4\pi}\log\frac{\Lambda^2}{M_W^2}\mathrm{Im}\Big\{6(c_T+c_T')-(c_S+c_S')-(c_P+c_P')\Big\}$$

$$\times\,[\bar{e}\mathrm{i}\gamma_5 e\,\bar{u}u\,+\,\bar{e}e\,\bar{u}\,\mathrm{i}\,\gamma_5 u\,+\,\frac{1}{2}\bar{e}\,\mathrm{i}\,\gamma_5\sigma_{\mu\nu}e\,\bar{u}\,\sigma_{\mu\nu}u]. \qquad (11.33)$$

The corresponding effective operator for the interaction between an electron and a d quark is

$$\frac{G}{\sqrt{2}}\frac{\alpha}{4\pi}\log\frac{\Lambda^2}{M_W^2}\mathrm{Im}\Big\{(c_S+c_S')-(c_P+c_P')\Big\}[\bar{e}\,\mathrm{i}\,\gamma_5 e\,\bar{d}d\,-\,\bar{e}\,e\,\bar{d}\,\mathrm{i}\,\gamma_5 d]. \qquad (11.34)$$

In all our calculations we assume for the W boson propagator the simple Feynman form $\delta_{\mu\nu}/(q^2-M_W^2)$ with the hope that the term $-q_\mu q_\nu/M_W^2$ in the numerator will somehow cancel out with higher accuracy calculations.

Now, the effective operators (11.33) and (11.34) generate the following CP-odd electron–quark constants (see Sect. 8.3.4):

$$k_{1u}=k_{3u}=2k_{2u}$$

$$=\frac{\alpha}{4\pi}\log\frac{\Lambda^2}{M_W^2}\mathrm{Im}\,\{6(c_T+c_T')-(c_S+c_S')-(c_P+c_P')\}\,; \qquad (11.35)$$

$$k_{1d}=-k_{3d}=\frac{\alpha}{4\pi}\log\frac{\Lambda^2}{M_W^2}\mathrm{Im}\,\{(c_S+c_S')-(c_P+c_P')\}\,. \qquad (11.36)$$

All these constants k are constrained by the atomic EDM experiments at the level of 10^{-7}. Therefore, the upper limits on the CP-odd β decay parameters are close to

$$\mathrm{Im}\,\{c_{S,T,P}+c_{S,T,P}'\}\,<\,10^{-4}. \qquad (11.37)$$

Now we have to pass over from the quark β decay constants to the usual ones that refer to nucleons. The scalar matrix element is given by formula (5.6):

$$\langle p|\bar{u}d|n\rangle\approx\bar{p}n.$$

The pseudoscalar matrix element is

$$\langle p|\bar{u}\mathrm{i}\gamma_5 d|n\rangle=\frac{f_\pi g_{\pi NN}\sqrt{2}}{m_u+m_d}\bar{p}\mathrm{i}\gamma_5 n, \qquad (11.38)$$

where $m_u=4$ MeV and $m_d=7$ MeV are the masses of the u and d quarks, $f_\pi=130$ MeV, $g_{\pi NN}=13.6$ is the strong πNN coupling constant. For the tensor matrix element, we use an order-of-magnitude estimate

$$\langle p|\bar{u}\sigma_{\mu\nu}d|n\rangle\approx\bar{p}\sigma_{\mu\nu}n. \qquad (11.39)$$

In this way we arrive at the following upper limits on the nucleon β decay constants:

$$\begin{aligned} &\mathrm{Im}\left\{C_{S,T} + C'_{S,T}\right\} < 10^{-4} \\ &\mathrm{Im}\left\{C_P + C'_P\right\} < 10^{-2}. \end{aligned} \tag{11.40}$$

The limits on diverse combinations of $\mathrm{Im}C_{S,T,P}$ obtained from direct experiments [364–368] lie roughly between 10^{-3} and unity, depending on the combination involved. All of those combinations differ, however, from ours. But what is more essential, our limits depend on some assumptions and are of a less quantitative character. Nevertheless, we believe that the constraints (11.40), being much more stringent than the direct upper limits on the corresponding constants, are quite interesting.

The transition to the two-loop approximation does not tell us anything interesting about the T-odd β decay interaction without derivatives. But it does say something about the so-called weak magnetism and weak dipole moment, i.e., about the β decay derivative coupling which is at the quark–lepton level

$$\frac{G}{\sqrt{2}}\frac{1}{2m_{\mathrm{p}}}\left[\bar{e}\gamma_\mu(1+\gamma_5)\nu\right]\left[\bar{u}(\tilde{g}_m + \tilde{g}_e\gamma_5)\sigma_{\mu\nu}k_\nu d\right] + \mathrm{h.c.} \tag{11.41}$$

The constants $\mathrm{Im}\{\tilde{g}_{m,e}\}$ describing the T-odd part of this interaction are close analogues of the dimensionless parameters q introduced above. It is no wonder therefore that in the two-loop approximation we get here the same upper limits

$$\mathrm{Im}\,\tilde{g}_{m,e} < 10^{-4}. \tag{11.42}$$

Dimensional considerations show that the same limits as (11.42) for the quark–lepton β decay constants also hold for the parameters of the nucleon weak current induced in this way:

$$\mathrm{Im}\,g_{m,e} < 10^{-4}. \tag{11.43}$$

The experimental proposal [369] aims at an accuracy $\approx 10^{-2}$ in the measurement of these constants.

* * *

With this note, we end this book. However, investigations of CP violation without strangeness – the searches for dipole moments – continue. The various upper limits set so far by the work we have described constitute a valuable contribution to elementary particle physics and to our knowledge of how Nature is arranged; the null results obtained so far are important. But in conclusion, we wish to confess that we dream of finding a positive result, of eventually discovering permanent electric dipole moments.

References

1. S.K. Lamoreaux: Practising the Poor Man's Brand of Particle Physics, Science **260**, 622 (1993)
2. J.H. Christensen, J.W. Cronin, V.L. Fitch, R. Turlay: Phys. Rev. Lett. **13**, 138 (1964)
3. A.D. Sakharov: Pis'ma ZhETF **5**, 32 (1967) [Sov. Phys. JETP Lett. **5**, 24 (1967)]
4. E.M. Purcell, N.F. Ramsey: Phys. Rev. **78**, 807 (1950).
5. P.A.M. Dirac: Rev. Mod. Phys. **21**, 392 (1949)
6. J.H. Smith: Ph.D. thesis, Harvard University, 1951 (unpublished). J.H. Smith, E.M. Purcell, N.F. Ramsey: Phys. Rev. **108**, 120 (1957)
7. T.D. Lee, C.N. Yang: Phys. Rev. **104**, 254 (1956)
8. N.F. Ramsey: In *Discovery of Weak Neutral Currents: The Weak Interaction Before and After*, eds. A.K. Mann, D.B. Cline (AIP, New York 1993)
9. C.S. Wu. E. Ambler, R. Hayward, D. Hoppes, R. Hudson: Phys. Rev. **105**, 1413 (1957)
10. R. Garwin, L. Lederman, M. Weinrich: Phys. Rev. **105**, 1415 (1957)
11. J. Friedman, V. Telegdi: Phys. Rev. **105**, 1681 (1957)
12. L.D. Landau: Zh. Eksp. Theor. Fiz. **32**, 405 (1957) [Sov. Phys. JETP **5**, 405 (1957)]; Nucl. Phys. **3**, 127 (1957)
13. J.D. Jackson, S.B. Treiman, H. Wyld, Jr.: Phys. Rev. **106**, 517 (1957)
14. N.F. Ramsey: Phys. Rev. **109**, 222 (1958)
15. L.D. Landau, E.M. Lifshitz: *Quantum Mechanics* (Nauka, Moscow 1989)
16. R.G. Sachs: *The Physics of Time Reversal* (University of Chicago, Chicago 1987)
17. Ya.B. Zel'dovich: Zh. Eksp. Theor. Fiz. **39**, 1483 (1960) [Sov. Phys. JETP **5**, 336 (1961)]
18. J.S. Bell: Nuovo Cim. **24**, 452 (1962)
19. A.M. Perelomov: Dokl. AN SSSR **146**, 75 (1962) [Sov. Phys. Dokl. **7**, 809 (1963)]
20. A.I. Baz', Ya.B. Zel'dovich, A.M. Perelomov: *Scattering, Reactions and Decays in Non-Relativistic Quantum Mechanics* (Israel Program for Scientific Translations, Jerusalem 1969)
21. J.S. Bell, G. Karl: Nuovo Cim. A **41**, 487 (1977)
22. G. Karl: Kvant.Elektr. **5**, 1743 (1978) [Sov. J. Quant. Electronics **5** (1978)]
23. I.B. Khriplovich: *Parity Nonconservation in Atomic Phenomena* (Gordon and Breach, London 1991)
24. Ya.A. Azimov, A.A. Ansel'm, A.N. Moskalev, R.M. Ryndin: Zh. Eksp. Theor. Fiz. **67**, 17 (1974) [Sov. Phys. JETP **33**, (1974)]
25. R.W. Dunford, R.R. Lewis, W.L. Williams: Phys. Rev. A **18**, 2421 (1978)

25a. W. Bernreuther, O. Nachtmann: Z. Phys. A **309**, 197 (1983)

25b. G.W. Botz, D. Bruss, O. Nachtmann: Ann. Phys. **240**, 107 (1995)

26. L. Stodolsky: Nucl. Phys. B **197**, 213 (1982)
27. I.B. Khriplovich, M.E. Pospelov: Z.Phys. D **17**, 81 (1990)
28. J. Schwinger: Phys. Rev. **91**, 720, 723 (1953); **94**, 1366 (1953)
29. G. Lüders: Kgl. Danske Videnskab Selskab. Mat. Fiz. Medd. **28**, No 5 (1954)
30. W. Pauli: In *Niels Bohr and the Development of Physics* (Pergamon Press, London 1955)
31. V.B. Berestetskii, E.M. Lifshitz, L.P. Pitaevskii: *Quantum Electrodynamics* (Pergamon Press, Oxford 1982)
32. M.S. Noecker, B.P. Masterson, C.E. Wieman: Phys. Rev. Lett. **61**, 310 (1988)
33. M.J.D. Macpherson, K.R. Zetie, R.B. Warrington, D.N. Stacey, J.P. Hoare: Phys. Rev. Lett. **67**, 2784 (1991)
34. N.H. Edwards, S.J. Phipp, P.E.G. Baird, S. Nakayama: Phys. Rev. Lett. **74**, 2654 (1995)
35. P.A. Vetter, D.M. Meekhof, P.K. Majumder, S.K. Lamoreaux, E. N. Fortson: Phys. Rev. Lett. **74**, 2658 (1995)
36. V.F. Dmitriev, I.B. Khriplovich, V.B. Telitsin: Phys. Rev. C **52**, 171 (1995)
37. E.R. Boston, P.G.H. Sandars: J. Phys. **B23**, 2663 (1990)
38. S.G. Porsev: Phys. Rev. A **49**, 5105 (1994)
39. A.D. Dolgov: Pis'ma Zh. Eksp. Theor. Fiz. **2**, 494 (1965) [Sov. Phys. JETP Lett. **2**, 308 (1965)]
40. F. Boudjema, C. Hamzaoui: Phys. Rev. D **43**, 3748 (1995)
41. I.Yu. Kobzarev, L.B. Okun', M.V. Terentjev: Pis'ma Zh. Eksp. Teor. Fiz. **2**, 466 (1965) (Sov. Phys. JETP Lett. **2**, 289 (1965))
42. I.B. Khriplovich, A.I. Vainshtein: unpublished
43. N.F. Ramsey: Rep. Prog. Phys. **45**, 95 (1982)
44. L.R. Hunter: Science **252**, 73 (1991)
45. *Time Reversal – The Art Rich Memorial Symposium*, eds. M. Skalsey, P.H. Bucksbaum, R.S. Conti, D.W. Gidley (AIP, New York 1993)
46. S.K. Lamoreaux, R. Golub: Phys. Rev. D **50**, 5632 (1994)
47. C. Cohen-Tannoudji, B. Diu, F. Laloë: *Quantum Mechanics* (Wiley, New York 1977)
48. R.P. Feynman, F.L. Vernon, R.W. Hellwarth: J. Appl. Phys. **28**, 49 (1957)
49. M. Pendlebury: Physics World, **2**, 55 (1989)
50. N.F. Ramsey: Physics Today, Oct. 1993, p. 40
51. N.F. Ramsey: *Molecular Beams* (OUP, London 1956)
52. L.C. Balling: Adv. Quant. Elec. **3**, 1 (1975)
53. W. Happer: Rev. Mod. Phys. **44**, 169 (1972)
54. H.G. Dehmelt: Phys. Rev. **105**, 1487 (1957)
55. W.T. Diamond: In *Proceedings of the 1993 Particle Accelerator Conference* (IEEE, New York 1993); R.J. Noer, Appl. Phys. A bf 28, 1 (1982); a rather complete bibliography is given by R. Hawley, A. Maitland, *Vacuum as an Insulator* (Chapman and Hall, London 1967)
56. J.G. Trump, R.G. van de Graaff: J. Appl. Phys. **18**, 327 (1947)
57. L.L. Alston. ed.: *High Voltage Technology* (OUP, London 1968)
58. R. Golub: Sov. Phys. Tech. Phys. **31**, 945 (1986)
59. J. Byrne et al.: Phys. Rev. Lett. **65**, 289 (1993)
60. J.M. Meek, J.D. Craggs: *Electric Discharges through Gases* (OUP, London 1953)
61. M. Forte: Revue Phys. Appl. **4**, 241 (1969)
62. C. Kittel: *Introduction to Solid State Physics* (Wiley, New York 1971)
63. S.K. Lamoreaux: Nucl. Instr. Meth. A **284**, 43 (1989)
64. T.J. Sumner, J.M. Pendlebury, K.F. Smith: J. Phys. D. **20**, 1095 (1987)
65. T.J. Sumner, Ph.D. Thesis, University of Sussex (1979) (unpublished)

66. J.D. Jackson: *Classical Electrodynamics*, 2nd. ed. (Wiley, New York 1975)
67. S.K. Lamoreaux: Report SKL-UW-1986-04 (unpublished)
68. R.J. Hanson, F.M. Pipkin: Rev. Sci. Inst. **36**, 179 (1964)
69. R. Golub: private communication (1996)
70. J.A. Stratton: *Electromagnetic Theory* (McGraw-Hill, New York 1941)
71. R. Golub: private communication (1987)
72. B.R. Heckel: private communication (1985)
73. F.E. Terman: *Radio Engineers' Handbook* (McGraw-Hill, New York 1943)
74. P.G.H. Sandars: E. Lipworth: Phys. Rev. Lett. **13**, 718 (1964)
75. E.D. Commins, S.B. Ross, D. DeMille, B.C. Regan: Phys. Rev. A **50**, 2960 (1994)
76. A. Abragam: *Principles of Nuclear Magnetism* (OUP, London 1962)
77. W. Happer: Phys. Rev. B **1**, 2203 (1970)
78. K.F. Smith et al.: Phys. Lett. B **234**, 191 (1990)
79. I.S. Altarev et al.: Phys. Lett. B **276**, 242 (1992)
80. J.P. Jacobs, W.M. Klipstein, S.K. Lamoreaux, B.R. Heckel, E.N. Fortson: Phys. Rev. A **52**, 3521 (1995); Phys. Rev. Lett. **71**, 3782 (1993)
81. R. Golub, S.K. Lamoreaux: Phys. Rep. **237**, 1 (1994)
82. R. Golub, J.M. Pendlebury: Contemp. Phys. **13**, 519 (1972)
83. G.G. Havens, I.I. Rabi, J. Rainwater: Phys. Rev. **72**, 637 (1947)
84. E. Fermi, L. Marshall: Phys. Rev. 72, 1139 (1947)
85. E. Amaldi: *Handbuch der Physik*, Bd. XXXVIII/2 (Springer, Berlin 1959)
86. J. Chadwick, M. Goldhaber: Nature **134**, 237 (1934); Proc. Roy. Soc. A **151**, 479 (1935)
87. A.H. Snell, L.C. Miller: Phys. Rev. **74**, 1217 (1948)
88. E. Fermi: Phys. Rev. **48**, 570 (1935)
89. Particle Data Group: Phys. Rev. D **50**, 1173 (1994)
90. E. Fermi, L. Marshall: Phys. Rev. **71**, 666 (1947)
91. V.F. Sears: *Neutron Optics* (OUP, London 1989)
92. V.F. Turchin: *Cold Neutron* (Program for Scientific Translations, Israel 1965)
93. L.L. Foldy: Phys. Rev. **67**, 107 (1945)
94. E. Fermi: *Collected Papers* (University of Chicago, Chicago 1962) pp. 639-641
95. H. Maier-Leibnitz, T. Springer: J. Nucl. Energy A/B **17**, 217 (1963)
96. L.W. Alvarez, F. Bloch: Phys. Rev. **57**, 111 (1940)
97. P.D. Miller, W.B.Dress, J.K. Baird, N.F. Ramsey: Phys. Rev. Lett. **19**, 381, (1967)
98. C.G. Shull, R. Nathans: Phys. Rev. Lett. **19**, 384 (1967)
99. C.G. Shull: Phys. Rev. Lett. **10**, 297 (1963)
100. M.L. Goldberger, F. Seitz: Phys. Rev. **71**, 294 (1947)
101. C.G. Shull: Phys. Rev. Lett. **21**, 1585 (1968)
102. M. Forte, C.M.E. Zeyen: Nucl. Instr. and Meth. A **284**, 147 (1989)
103. V.L. Alexeev et al.: Nucl. Instr. and Meth. A **284**, 181 (1989)
104. V.V. Fedorov, V.V. Voronin, E.G. Lapin: In *Work Meeting of the High Flux Reactor PIK Project* (Petersburg Nucl. Phys. Inst., 1992, unpublished); Tech. Phys. Lett. **21**, 881 (1995); *ibid*, p. 884.
105. W.B. Dress, J.K. Baird, P.D. Miller, N.F. Ramsey: Phys. Rev. **170**, 1200 (1968)
106. V.W. Cohen et al.: Phys. Rev. **177**, 1942 (1969)
107. J.K. Baird et al.: Phys. Rev. **179**, 1285 (1969)
108. S. Apostolescu et al.: Rev. Roum. Phys **15**, 343 (1070)
109. W.B. Dress, P.D. Miller, N.F. Ramsey: Phys. Rev. D **7**, 3147 (1973)
110. W.B. Dress et al.: Phys. Rev. D **15**, 9 (1977)

111. Ya. B. Zel'dovich: Zh. Eksp. Theor. Fiz. **36**, 1952 (1959) [Sov. Phys. JETP **9**, 1389 (1959)]
112. R. Golub, D.J. Richardson, S.K. Lamoreaux: *Ultracold Neutrons* (Adam Hilger, Bristol 1991)
113. V.K. Ignatovich: *The Physics of Ultracold Neutrons* (OUP, London 1990)
114. J.M. Pendlebury: In *15th Nuclear Physics Divisional Conference. Low Energy Nuclear Dynamics. St. Petersburg, Russia* (World Scientific, Singapore 1995)
115. V.I. Lushchikov, Yu.N. Pokotilovsky, A.V. Strelkov, F.L. Shapiro: Pis'ma ZhETF **9**, 40 (1969) [Sov. Phys. JETP Lett. **9**, 23 (1969)]
116. A. Steyerl: Phys. Lett. B 29, 33 (1969)
117. F.L. Shapiro: Usp. Fiz. Nauk **95**, 145 (1968) [Sov. Phys. Uspekhi **11**, 345 (1968)]
118. L.B. Okun': Comm. Nucl. Part. Phys. **3**, 133 (1969)
119. P.C. Miranda: Ph.D. Thesis (University of Sussex, 1987) (unpublished)
120. J.M. Pendlebury et al.: Phys. Lett. B **136**, 327 (1984)
121. A. Steyerl et al.: Phys. Lett. A **116**, 347 (1986)
122. K.F. Smith et al.: Phys. Lett. B **234**, 191 (1990)
123. W. Lanford, R. Golub: Phys. Rev. Lett. **39**, 1509 (1977)
124. W. Mampe, P. Ageron, R. Gähler: Z. Phys. B **45**, 1 (1981)
125. P.R. Bevington: *Data Reduction and Error Analysis for the Physical Sciences* (McGraw-Hill, New York 1969)
126. J.M. Pendlebury: in *Proc. Ninth Symposium on Grand Unification* (Aix-les-Bains, France 1988)
127. I.S. Altarev et al.: Pis'ma ZhETF **44**, 344 (1986) [Sov. Phys. JETP Lett. **44**, 460 (1981)]
128. I.S. Altarev et al.: Nucl. Phys. A **341**, 269 (1980)
129. I.S. Altarev et al.: Phys. Lett. B **102**, 13 (1981)
130. I.S. Altarev et. al: Phys. Lett. B **276**, 242 (1992)
131. E.B. Aleksandrov et al.: Opt. i Spektr. **78**, 325 (1995) [Sov. Phys. Opt. Spectr. **78**, 292 (1995)]
132. N.F. Ramsey: Acta Physica Hungarica **55**, 117 (1984)
133. J.M. Pendlebury: Nucl. Phys. A **546**, 359c (1992)
134. S.K. Lamoreaux: Ph.D. Thesis (University of Washington, 1986) (unpublished); S.K. Lamoreaux et al., Phys. Rev. A **39**, 1082 (1989)
135. R. Golub, J.M. Pendlebury: Phys. Lett. A **62**, 3376 (1977)
136. P. Ageron et al.: Phys. Lett. A **66**, 469 (1978)
137. R. Golub et al.: Z. Phys. B **51**, 187 (1983)
138. S.K. Lamoreaux, R. Golub: Pis'ma ZhETF **58**, 844 (1995) [Sov. Phys. JETP Lett. **58**, 792 (1993)]
139. R. Golub: Phys. Lett. A **72**, 387 (1979)
140. S.K. Lamoreaux: J. Neutron Research **4**, 183 (1996)
141. S.K. Lamoreaux: In *Proc. of the Baryon Instability Conf., Report ORNL–6910* (Oak Ridge, TN 1996)
142. P.A. Egelstaff, R.S. Pease: J. Sci. Instr. **31**, 207 (1954)
143. V.V. Vasil'yev et al.: Institute of Theoretical and Experimental Physics Report 9-95 (Moscow 1995) (unpublished)
144. H. Yoshiki et al.: Phys. Rev. Lett. **68**, 1323 (1992); R. Golub, S.K. Lamoreaux: Phys. Rev. Lett. **70**, 517 (1993)
145. A.I. Kilvington et al.: Phys. Lett. A **125**, 416 (1987)
146. L. Passell, R. Schermer: Phys. Rev. **150**, 146 (1960)
147. C.G. Aminoff et al.: Rev. Phys. Appl. **24**, 827 (1989)
148. J.M. Doyle, S.K. Lamoreaux: Europhys. Lett. **26**, 253 (1994)

149. R. Golub: J.Physique **44**, L321 (1983); *Proc. 18th Inter.Conf. on Low-Temperature Physics, Part 3; Invited Papers* (Kyoto 1987) p.2073.
150. L.B. Okun': *Leptons and Quarks* (North Holland, Amsterdam 1982)
151. L.D. Landau, E.M. Lifshitz: *The Classical Theory of Fields* (Nauka, Moscow 1988)
152. V. Baluni: Phys. Rev. D **19**, 2227 (1979)
153. R.J. Crewther, P. Di Vecchia, G. Veneziano, E. Witten: Phys. Lett. B **88**, 123 (1979); **91**, 487(E) (1980)
154. J. Ellis, M.K. Gaillard: Nucl. Phys. B **150**, 141 (1979)
155. I.B. Khriplovich, A.I. Vainshtein: Nucl. Phys. B **414**, 27 (1994)
156. R.D. Peccei, H.R. Quinn: Phys. Rev. Lett. **38**, 1440 (1977); Phys. Rev. D **16**, 1791 (1977)
157. S. Weinberg: Phys. Rev. Lett. **40**, 223 (1978)
158. F. Wilczek: Phys. Rev. Lett. **40**, 279 (1978)
159. S. Weinberg: Phys. Rev. Lett. **63**, 2333 (1989)
160. I.I. Bigi, N.G. Uraltsev: Nucl. Phys. B **353**, 321 (1991)
161. M. Kobayashi, T. Maskawa: Prog. Theor. Phys. **49**, 652 (1973)
162. L. Maiani: Phys. Lett. B **62**, 183 (1976)
163. J. Ellis, M.K. Gaillard, D.V. Nanopoulos: Nucl. Phys. B **109**, 23 (1976)
164. B. Lee: Phys. Rev. D **15**, 3394 (1977)
165. S.L. Glashow, J. Iliopoulos, L. Maiani: Phys. Rev. D **2**, 1285 (1970)
166. E.P. Shabalin: Yad. Fiz. **28**, 151 (1978) [Sov. J. Nucl. Phys. **28**, 75 (1978)]
167. M.E. Pospelov, I.B. Khriplovich: Yad. Fiz. **53**, 1030 (1991) [Sov. J. Nucl. Phys. **53**, 638 (1991)]
168. I.B. Khriplovich, M.E. Pospelov: Nucl. Phys. B **420**, 505 (1994)
169. I.B. Khriplovich: Phys. Lett. B **173**, 193 (1986); Yad. Fiz. **44**, 1019 (1986) [Sov. J. Nucl. Phys. **44**, 659 (1986)]
170. I.B. Khriplovich, A.R. Zhitnitsky: Phys. Lett. B **109**, 490 (1982)
171. M.B. Gavela, A. Le Yaouanc, L. Oliver, O. Pène, J.-C. Raynal, T.N. Pham: Phys. Lett. B **109**, 215 (1982)
172. C. Hamzaoui, M.E. Pospelov: Phys. Lett. B **357**, 616 (1995)
173. C. Hamzaoui, M.E. Pospelov: Phys. Rev. D **54**, 2194 (1996)
174. T.D. Lee, Phys. Rev. D **8**, 1226 (1973)
175. S. Weinberg: Phys. Rev. Lett. **37**, 657 (1976)
176. A.A. Anselm, D.I. Dyakonov: Nucl. Phys. B **145**,271 (1978)
177. A.A. Anselm, N.G. Uraltsev: Yad. Fiz. **30**, 465 (1979) [Sov. J. Nucl. Phys. **30** (1979)]
178. A.R. Zhitnitsky, I.B. Khriplovich: Yad. Fiz. **34**, 167 (1981) [Sov. J. Nucl. Phys. **34** (1982)]
179. A.A. Anselm, V.E. Bunakov, V.P. Gudkov, N.G. Uraltsev: Pis'ma Zh. Eksp. Theor. Fiz. **40**, 310 (1984) [Sov. Phys. JETP Letters **40** (1984)]; Phys. Lett. B **152**, 116 (1985)
180. V.M. Khatsymovsky, I.B. Khriplovich, A.R. Zhitnitsky: Z. Phys. C **36**, 455 (1987)
181. S.M. Barr, A. Zee: Phys. Rev. Lett. **65**, 21 (1990)
182. J.F. Gunion, R. Vega: Phys. Lett. B **157**, 157 (1990)
183. R.G. Leigh, S. Paban, R.-M. Xu: Nucl. Phys. B **352**, 45 (1991)
184. G.F. Gunion, D. Wyler: Phys. Lett. B **248**, 170 (1990)
185. D. Chang, W.-Y. Keung, T.C. Yuan: Phys. Lett. B **251**, 608 (1990)
186. X.-G. He, B.H.J. McKellar, S. Pakvasa: Phys. Lett. B **254**, 231 (1991)
187. M.A. Shifman, A.I. Vainshtein, V.I. Zakharov: Phys. Rev. D **77**, 2583 (1978)
188. A. Manohar, H. Georgi: Nucl. Phys. B **234**, 189 (1984)
189. H. Georgi, L. Randall: Nucl. Phys. B **276**, 241 (1986)

190. V.M. Khatsymovsky, I.B.Khriplovich: Phys. Lett. B **296**, 219 (1992)
191. I.B. Khriplovich: Phys. Lett. B **382**, 145 (1996)
192. X.-G. He, B.H.J. McKellar, S. Pakvasa: Int. J. Mod. Phys. A **4**, 5011 (1989)
193. S. Dimopoulos, L.J. Hall: Phys. Lett. B **344**, 185 (1995)
194. I.B. Khriplovich, K.N. Zyablyuk: Phys. Lett. B **383**, 429 (1996)
195. V.M. Khatsymovsky, I.B.Khriplovich, A.S. Yelkhovsky: Ann. Phys. **186**, 1 (1988)
196. X.-G. He, B.H.J. McKellar: Preprint hep-ph/9604394, 24 April 1996.
197. V.M. Khatsymovsky: Yad.Fiz. **53**, 548 (1991) [Sov.J.Nucl. Phys. **53** (1991)]
198. I.B. Khriplovich: unpublished
199. R.L. Garwin, L.M. Lederman: Nuovo Cim. **11**, 776 (1959)
200. L.I. Schiff: Phys. Rev. **132**, 2194 (1963)
201. I.B. Khriplovich: *Proc. of the XVII LIYaF Winter School* (Leningrad 1982)
202. E.E. Salpeter: Phys. Rev. **112**, 1642 (1958)
203. D.F. Nelson et al.: Phys. Rev. Lett. **2**, 492 (1959)
204. E.S. Ensberg: Bull. Am. Phys. Soc. **7**, 543 (1962); Phys. Rev. **153**, 36 (1967); H.G. Robinson, E.S. Ensberg, H.G. Dehmelt, Bull. Am. Phys. Soc. **3**, 9 (1958)
205. P.G.H. Sandars, E. Lipworth: Phys. Rev. Lett. **13**, 718 (1964); *ibid*, p. 715
206. P.G.H. Sandars: In *Time Reversal – the Arthur Rich Memorial Symposium*, eds. M. Skalsey, P. Bucksbaum, R.S. Conti, D.W. Gidley (AIP, New York 1991)
207. P.G.H. Sandars: Phys. Lett. **14**, 194 (1965)
208. V.V. Flambaum: Yad. Fiz. **24**, 383 (1976) [Sov. J. Nucl. Phys. **24**, (1976)]
209. P.G.H. Sandars: Phys. Lett. **22**, 290 (1966); J. Phys. B **1**, 511 (1968)
210. M.A. Player, P.G.H. Sandars: J. Phys. B **3**, 1620 (1970)
211. A.C. Hartley, E. Lindroth, A.-M. Mårtensson-Pendrill: J. Phys. B **23**, 3417 (1990)
212. Z.W. Liu, H.P. Kelly: Phys. Rev. A **45**, R4210 (1992)
213. R.E. Rand: Phys. Rev. **140**, B1605 (1965)
214. M.C. Weisskopf et al.: Phys. Rev. Lett. **21**, 1645 (1968)
215. H. Gould: Phys. Rev. Lett. **24**, 1091 (1970); see also J.P. Carrico et al.: Phys. Rev. A **1**, 211 (1970)
216. B.V. Vasil'iev, E.V. Kolycheva: Zh. Eksp. Theor. Fiz. **74**, 466 (1978) [Sov. Phys. JETP **47**, 243 (1978)]
217. F.R. Huang-Hellinger: Ph.D. Thesis (University of Washington, Seattle, WA 1987) (unpublished)
218. S.A. Murthy, D. Krause Jr., Z.L. Li, L.R. Hunter: Phys. Rev. Lett. **63**, 965 (1989)
219. K. Abdullah et al.: Phys. Rev. Lett. **65**, 2347 (1990)
220. E.D. Commins, S.B. Ross, D. DeMille, B.C. Regan: Phys. Rev. A **50**, 2960 (1994)
221. Chen Xu, F.R. Huang-Hellinger, B.R. Heckel, E.N. Fortson: Phys. Rev. A **50**, 4729 (1996)
222. S.K. Lamoreaux, E.N. Fortson: Phys. Rev. A **46**, 7053 (1992)
223. L.R. Hunter: private communication (1996)
224. E.D. Commins: D.P. DeMille, private communication (1995)
225. M. Bijlsma, B.J. Verhaar, D.J. Heinzen: Phys. Rev. A **49**, R4285 (1994)
226. K. Gibble, S. Chu: Phys. Rev. Lett. **70**, 1711 (1993)
227. W.M. Klipstein: private communication (1996)
228. D. Budker, D.P. DeMille: private communication (1996)
229. A. Weis: In *Proc. of the XXX Recontre de Moriond* eds. B. Guiderdoni, G. Greene, E.A. Hinds, J. Tran Thanh Van (Edition Frontieres, Gif-sur-Yvette, France 1995)

230. S.K. Lamoreaux, I.B. Khriplovich: unpublished
231. R.M. Sternheimer: Phys. Rev. **113**, 828 (1959)
232. B. Rose: In *Proceedings of the 10th International Conference on High Energy Physics* (New York 1960) p.787
233. C.L. Cowan, F. Reines: Phys. Rev. **107**, 528 (1957)
234. S. Rosendorff: Nuovo Cim. **27**, 251 (1960)
235. J. Bailey et al.: Phys. Lett. B **68**, 191 (1977)
236. F.J.M. Farley et al.: J. Phys. G **4**, 345 (1978)
237. W.K. Gibson, K. Green: Nuovo Cim. A **45**, 882 (1966)
238. L. Pondrom et al.: Phys. Rev. D **23**, 814 (1981)
239. F. del Aguila, M. Sher: Phys. Lett. B **252**, 116 (1990)
240. J.A. Grifols, A. Méndez: Phys. Lett. B **255**, 611 (1991)
241. R. Escribano, E. Massó: Phys. Lett. B **301**, 419 (1993)
242. T.G. Vold, F.J. Raab, B.R. Heckel, E.N. Fortson: Phys. Rev. Lett. **52**, 2229 (1984)
243. B. Grover: Phys. Rev. Lett. **40**, 391 (1978)
244. C.H. Volk, T.M. Kwon, J.G. Mark: Phys. Rev. A **21**, 1549 (1980)
245. N.D. Bhaskar, W. Happer, M. Larsson, X. Zeng: Phys. Rev. Lett. **50**, 105 (1983); W. Happer et al.: Phys. Rev. A **29**, 3092 (1984)
246. C.C. Bouchiat, M.A. Bouchiat, L.C. Pottier: Phys. Rev. **181**, 3144 (1969)
247. S.K. Lamoreaux: Ph.D. Thesis (University of Washington, Seattle, WA 1986) (unpublished)
248. S.K. Lamoreaux et al.: Phys. Rev. Lett. **59**, 2275 (1987)
249. J.P. Jacobs: Ph.D. Thesis (University of Washington, Seattle, WA 1992) (unpublished)
250. W.M. Klipstein: Ph.D. Thesis (University of Washington, Seattle, WA 1996) (unpublished)
251. X. Zeng et al.: Phys. Rev. A **31**, 260 (1985)
252. T.E. Chupp: In *Time Reversal – The Arthur Rich Memorial Symposium* eds. M. Skalsey, P. Bucksbaum, R.S. Conti, D.W. Gidley (AIP, New York 1991)
253. H.G. Robinson, T. Myint: Appl.Phys. Lett. **5**, 116 (1964)
254. T.E. Chupp, R.J. Hoare, R.L. Walsworth, Bo Wu: Phys. Rev. Lett. **72**, 2363 (1994)
255. Y. Chibane, S.K. Lamoreaux, J.M. Pendlebury, K.F. Smith: Meas. Sci. and Tech. **6**, 1671 (1995)
256. A.-M. Mårtensson-Pendrill: *Methods in Computational Chemistry, Volume 5: Atomic, Molecular Properties*, ed. S. Wilson (Plenum Press, New York 1992)
257. I.I. Gol'dman, V.D. Krivchenkov: *Collection of Problems on Quantum Mechanics* (GITTL, Moscow 1957)
258. I.I. Sobel'man: *Introduction to the Theory of Atomic Spectra* (Nauka, Moscow 1977)
259. C.E. Moore: *Atomic Energy Levels*, vol. 3, (NBS, Washington 1958)
260. L.N. Shabanova, Yu.N. Monakov, A.N. Khlyustalov: Opt. i Spektr. **47**, 3 (1979)
261. R.M. Sternheimer: Phys. Rev. **183**, 112 (1969)
262. V.K. Ignatovich: Zh. Eksp. Theor. Fiz. **56**, 2019 (1969) [Sov. Phys. JETP **29**, 1084 (1969)]
263. W.R. Johnson, D.S. Guo, M. Idrees, J. Sapirstein: Phys. Rev. A **32**, 2093 (1985)
264. W.R. Johnson, D.S. Guo, M. Idrees, J. Sapirstein: Phys. Rev. A **34**, 1043 (1986)
265. A.-M. Mårtensson-Pendrill, P. Öster: Phys. Scripta **36**, 444 (1987)

266. P.G.H. Sandars, R.M. Sternheimer: Phys. Rev. A **11**, 473 (1975)
267. A.Ya. Kraftmakher: J. Phys. B **21**, 2803 (1988)
268. R.M. Ryndin: private communication (cited in [23])
269. V.V. Flambaum: Phys. Lett. B **320**, 211 (1994)
270. P.G.H. Sandars: In *Atomic Physics 4, Proceedings of the 4th International Conference on Atomic Physics*, ed. G. zu Putlitz (Plenum, New York 1975)
271. C. Bouchiat: Phys. Lett. B **57**, 284 (1975)
272. E.A. Hinds, C.E. Loving, P.G.H. Sandars: Phys. Lett. B **62**, 97 (1976)
273. A.-M. Mårtensson-Pendrill: Phys. Rev. Lett. **54**, 1153 (1985)
274. V.A. Dzuba, V.V. Flambaum, P.G. Silvestrov: Phys. Lett. B **154**, 93 (1985)
275. V.V. Flambaum, I.B. Khriplovich: Zh. Eksp. Theor. Fiz. **89**, 1505 (1985) [Sov. Phys. JETP **62**, 872 (1985)]
276. M.G. Kozlov: Phys. Lett. A **130**, 426 (1988); private communication (1990)
277. E. Lindroth, A.-M. Mårtensson-Pendrill: Europhys. Lett. **15**, 155 (1991)
278. O.P. Sushkov, V.V. Flambaum, I.B. Khriplovich: Zh. Eksp. Theor. Fiz. **87**, 1521 (1984) [Sov. Phys. JETP **60**, 873 (1984)]
279. V.N. Novikov, O.P. Sushkov, V.V. Flambaum, I.B. Khriplovich: Zh. Eksp. Theor. Fiz. **73**, 802 (1977) [Sov. Phys. JETP **46**, 420 (1977)]
280. T.P. Cheng: Phys. Rev. D **13**, 2161 (1976)
281. J.F. Donoghue, Ch.R. Nappi: Phys. Lett. B **168**, 105 (1986)
282. D. DeMille: private communication (1996)
283. E.N. Fortson: Bull. Am. Phys. Soc. **28**, 1321 (1983)
284. H. Collard et al: Phys. Rev. Lett. **11**, 132 (1963)
285. H. Theissen, H. Fink, H.A. Bentz: Z. Phys. **231**, 475 (1970)
286. H. De Vries, G.J.G. Van Niftric, L. Lapikas: Phys. Lett. B **33**, 403 (1970)
287. I.B. Khriplovich: Zh. Eksp. Theor. Fiz. **71**, 51 (1976) [Sov. Phys. JETP **44**, 25 (1976)]
288. V.B. Telitsin: In *Proceedings of the 34th Conference on Nuclear Spectroscopy and Structure of Atomic Nucleus* (Nauka, Leningrad 1984) p.190
289. V.V. Flambaum, I.B. Khriplovich, O.P. Sushkov: Phys. Lett. B **162**, 213 (1985); Nucl. Phys. A **449**, 750 (1986)
290. I.B. Khriplovich: In *The First Arctic Workshop on Future Physics and Accelerators*, eds. M. Chaichian, K. Huitu, R. Orava (World Scientific, Singapore 1995) p.217
291. P.G.H. Sandars: Phys. Rev. Lett. **19**, 1396 (1967)
292. O.P. Sushkov, V.V. Flambaum: Zh. Eksp. Theor. Fiz. **75**, 1208 (1978) [Sov. Phys. JETP **48**, 608 (1978)]
293. V.G. Gorshkov, L.N. Labzowsky, A.N. Moskalev: Zh. Eksp. Theor. Fiz. **76**, 414 (1979) [Sov. Phys. JETP **49**, 209 (1979)]
294. V.A. Onishchuk: Preprint JINR P4-3299/Dubna (1967)
295. G.E. Harrison, P.G.H. Sandars, S.J. Wright: Phys. Rev. Lett. **22**, 1263 (1969)
296. E.A. Hinds, P.G.H. Sandars: Phys. Rev. A **21**, 480 (1980)
297. D.A Wilkening, N.F. Ramsey: Phys. Rev. A **29**, 425 (1984)
298. D. Schropp, D. Cho, T. Vold, E.A. Hinds: Phys. Rev. Lett. **59**, 991 (1987)
299. D. Cho, K. Sangster, E.A. Hinds: Phys. Rev. Lett. **63**, 2559 (1989)
300. D. Cho, K. Sangster, E.A. Hinds: Phys. Rev. A **44**, 2783 (1991)
301. H. Dijkerman et al.: Z. Naturforch. A **27**, 425 (1972)
302. B.N. Ashkinadzi et al.: PNPI Preprint No.1801 (Petersburg Nucl. Phys. Inst., Gatchina, Russia 1992, unpublished); V.F. Ezhov et al., Tech. Phys. Lett. **21**, 598 (1992)
303. E.A. Hinds, P.G.H. Sandars: Phys. Rev. A **21**, 471 (1980)
304. P.V. Coveney, P.G.H. Sandars: J. Phys. B **16**, 3727 (1983)
305. G. Gräf, W. Paul, C.H. Schlier: Z.Phys. **153**, 38 (1958)

306. P.V. Coveney: Part II BA thesis, Oxford University, 1981 (unpublished)
307. E.A. Hinds, K. Sangster: In *Time Reversal – The Arthur Rich Memorial Symposium*, eds. M. Skalsey, P. Bucksbaum, R.S. Conti, D.W. Gidley (AIP, New York 1991) (Some misprints in [304] are pointed out here)
308. B.E. Sauer, Jun Wang, E.A. Hinds: Phys. Rev. Lett. **74**, 1554 (1995)
309. M.G. Kozlov, V.F. Ezhov: Phys. Rev. A **49**, 4502 (1994)
310. W.J. Childs, G.L. Goodman, L.S. Goodman: J. Mol. Spectr. **86**, 365 (1981)
311. H. Lefebvre-Brion, R.W. Field: *Perturbations in the Spectra of Diatomic Molecules* (Academic Press, New York 1986)
312. W.C. Martin, R. Zalubas, L. Hagan: *Atomic Energy Levels - The Rare-Earth Elements* (NBS, Washington 1978)
313. I.B. Khriplovich, unpublished
314. K.L. Dunfield et al.: J. Mol. Spectr. **174**, 433 (1995)
315. M. Dolg, H. Stoll, H. Preuss: Chem. Phys. **165**, 21 (1992)
316. V.V. Flambaum, I.B. Khriplovich: Phys. Lett. A **110**, 121 (1985)
317. M.G. Kozlov: Zh. Eksp. Theor. Fiz. **89**, 1933 (1985) [Sov. Phys. JETP **62**, 1114 (1985)]
318. M.G. Kozlov et al.: J. Phys. B **20**, 4939 (1987)
319. Yu.Yu. Dmitriev et al.: Phys. Lett. A **167**, 280 (1992)
320. M.G. Kozlov, L.N. Labzowsky: J. Phys. B **28**, 1933 (1995)
321. A.B. Migdal: *Theory of Finite Fermi Systems and Applications to Atomic Nuclei* (Interscience, New York 1967)
322. V.V. Flambaum: In *Time Reversal Invariance and Parity Violation in Neutron Reactions*, ed. C.R. Gould, J.D. Bowman and Yu.P. Popov (World Scientific, Singapore 1994)
323. V.V. Flambaum, O.K. Vorov: Phys. Rev. C **51**, 1521 (1995)
324. V.V. Flambaum: O.K. Vorov: Phys. Rev. C **51**, 2914 (1995)
325. I.B. Khriplovich: Phys. Rev. C **52**, 1132 (1995)
326. A. Bohr, B.R. Mottelson: *Nuclear Structure* (Benjamin, New York 1969)
326a. V.V. Flambaum, O.P. Sushkov: private communication (1984)
327. V.F. Dmitriev, I.B. Khriplovich, V.B. Telitsin: Phys. Rev. C **50**, 2358 (1994)
327a. V.F. Dmitriev, V.B. Telitsin, V.V. Flambaum, V.A. Dzuba: Phys. Rev. C **54**, 1 (1996)
328. G. Feinberg: Trans. N.Y. Acad. Sci. **38**, 26 (1977)
329. W.C. Haxton, E.M. Henley: Phys. Rev. Lett. **51**, 1937 (1983)
330. N. Auerbach, V.V. Flambaum, V. Spevak: Phys. Rev. Lett. **76**, 4316 (1996)
331. I. Akhmad, P.A. Butler: Annu. Rev. Nucl. Part. Sci. **43**, 71 (1993)
332. J.F. Donoghue, B.R.Holstein, M.J. Musolf: Phys. Lett. B **196**, 196 (1987)
333. X.-G. He, B.H.J. McKellar: Phys. Rev. D **46**, 2131 (1992)
334. A.I. Vainshtein, V.I. Zakharov, M.A. Shifman: Zh. Eksp. Theor. Fiz. **72**, 1275 (1977) [Sov. Phys. JETP **45**, (1977)]
335. I.B. Khriplovich: unpublished
336. N.K. Cheung, H.E. Henrikson, F. Boehm: Phys. Rev. C **16**, 2381 (1977)
337. J. Bystricky, F. Leah, P. Winternitz: J. Phys. **45**, 207 (1984)
338. C.A. Davies et al.: Phys. Rev. C **33**, 1196 (1986)
339. J.B. French, A. Pandey, J. Smith: In *Tests of Time Reversal Invariance in Neutron Physics*, ed. N.R. Roberson, C.R. Gould, J.D. Bowman (World Scientific, Singapore 1987)
340. M.G. Kozlov, S.G. Porsev: Phys. Lett. A **142**, 233 (1989); Yad. Fiz. **51**, 1056 (1990) [Sov. J. Nucl. Phys. **51** (1990)]
341. A.N. Moskalev, S.G. Porsev: Yad. Fiz. **49**, 1266 (1989) [Sov. J. Nucl. Phys. **49** (1989)]

342. R.S. Conti: In *Time Reversal – The Arthur Rich Memorial Symposium*, eds. M. Skalsey, P. Bucksbaum, R.S. Conti, D.W. Gidley (AIP, New York 1991)
343. R.S. Conti, S. Hatamian, A. Rich: Phys. Rev. A **33**, 3495 (1986)
344. P. Herczeg, J. Kambor, M. Simonius, D. Wyler: to be published
345. L. Wolfenstein, Nucl. Phys. B **77**, 375 (1974)
346. P. Herczeg: In *Tests of Time Reversal Invariance in Neutron Physics*, ed. N.R.Roberson, C.R. Gould, J.D. Bowman (World Scientific, Singapore 1987)
347. M.G. Kozlov: private communication (1990)
348. I.B. Khriplovich: unpublished
349. I.B. Khriplovich: Nucl. Phys. B **352**, 385 (1991)
350. E. Stephens: Ph.D. thesis, University of Oxford, 1992 (unpublished)
351. W.C. Haxton, A. Höring: Nucl. Phys. A **560**, 469 (1993)
352. W.C. Haxton, A. Höring, M. Musolf: Phys. Rev. D **50**, 3422 (1994)
353. J. Engel, C.R. Gould, V. Hnizdo: Phys. Rev. Lett. **73**, 3508 (1994)
354. O.K. Vorov: Phys. Lett. B **368**, 191 (1996)
355. J. Engel, P.H. Frampton, R.P. Springer: Phys. Rev. D **53**, 5112 (1996)
356. I.B. Khriplovich: In *Time Reversal – The Arthur Rich Memorial Symposium*, eds. M. Skalsey, P. Bucksbaum, R.S. Conti, D.W. Gidley (AIP, New York 1991)
357. I.B. Khriplovich: Pis'ma Zh. Eksp. Teor. Fiz. **52**, 1065 (1990) [Sov. Phys. JETP Lett. **52**, 461 (1990)]
358. R.S.Conti, I.B. Khriplovich: Phys. Rev. Lett. **68**, 3262 (1992)
359. E.C.G. Sudarshan: Proc. Roy. Soc. London A **305**, 319 (1968)
360. R. Brian, A. Gersten: Phys. Rev. Lett. **26**, 1000 (1971); **27**, 1102(E) (1971)
361. J. Binstock, R. Brian, A. Gersten: Phys. Lett. B **48**, 77 (1974)
362. M. Simonius, D. Wyler: Nucl. Phys. A **286**, 182 (1977)
363. M. Simonius: Phys. Lett. B **58**, 147 (1975)
364. R.I. Steinberg, P. Liaud, B. Vignon, V. Hughes: Phys. Rev. D **13**, 2469 (1976)
365. B.G. Erozolimsky, Yu.A. Mostovoi, V.P. Fedunin, A.I. Frank, O.V. Khakhan: Yad. Fiz. **28**, 98 (1978) [Sov. J .Nucl. Phys. **28**, 48 (1978)]
366. A.L. Hallin, F.P. Calaprice, D.W. MacArthur, L.E. Piilinen, M.B. Schneider, D.F. Schreiber: Phys. Rev. Lett. **52**, 337 (1984)
367. M.B. Schneider et al.: Phys. Rev. Lett. **51**, 1239 (1983)
368. M. Skalsey, M.S. Hatamian: Phys. Rev. C **31**, 2218 (1984)
369. L. De Braeckeleer, E.G. Adelberger: *Research Proposal. Precision Tests of Time-Reversal Invariance, Vector Current Conservation and Second Class Currents via $e-\nu-\alpha$ Angular Correlations in $A=8$ β-Decays* (University of Washington 1990)

Index

Springer and the environment

At Springer we firmly believe that an international science publisher has a special obligation to the environment, and our corporate policies consistently reflect this conviction.

We also expect our business partners – paper mills, printers, packaging manufacturers, etc. – to commit themselves to using materials and production processes that do not harm the environment. The paper in this book is made from low- or no-chlorine pulp and is acid free, in conformance with international standards for paper permanency.

Made in the USA
Monee, IL
13 September 2026

62973757R00138